Studies in Logic

Volume 72

Fathoming Formal Logic: Vol II

Semantics and Proof Theory for Predicate Logic

Volume 62
Argumentation and Reasoned Action. Proceedings of the 1st European Conference on Argumentation, Lisbon 2015. Volume I
Dima Mohammed and Marcin Lewiński, eds

Volume 63
Argumentation and Reasoned Action. Proceedings of the 1st European Conference on Argumentation, Lisbon 2015. Volume II
Dima Mohammed and Marcin Lewiński, eds

Volume 64
Logic of Questions in the Wild. Inferential Erotetic Logic in Information Seeking Dialogue Modelling
Paweł Łupkowski

Volume 65
Elementary Logic with Applications. A Procedural Perspective for Computer Scientists
D. M. Gabbay and O. T. Rodrigues

Volume 66
Logical Consequences. Theory and Applications: An Introduction.
Luis M. Augusto

Volume 67
Many-Valued Logics: A Mathematical and Computational Introduction
Luis M. Augusto

Volume 68
Argument Technologies: Theory, Analysis and Appplications
Floris Bex, Floriana Grasso, Nancy Green, Fabio Paglieri and Chris Reed, eds

Volume 69
Logic and Conditional Probability. A Synthesis
Philip Calabrese

Volume 70
Proceedings of the International Conference. Philosophy, Mathematics, Linguistics: Aspects of Interaction, 2012 (PhML-2012)
Oleg Prosorov, ed.

Volume 71
Fathoming Formal Logic: Volume I. Theory and Decision Procedures for Propositional Logic
Odysseus Makridis

Volume 72
Fathoming Formal Logic: Volume II. Semantics and Proof Theory for Predicate Logic
Odysseus Makridis

Studies in Logic Series Editor
Dov Gabbay dov.gabbay@kcl.ac.uk

Fathoming Formal Logic: Vol II

Semantics and Proof Theory for Predicate Logic

Odysseus Makridis

Fairleigh Dickinson University, NJ, USA

ISBN 978-1-84890-267-1

College Publications
Scientific Director: Dov Gabbay
Managing Director: Jane Spurr

http://www.collegepublications.co.uk

PREFACE

This text, volume II of a two-volume work titled *Fathoming Formal Logic*, examines in depth the so-called "standard" predicate logic. While the first volume of this work investigates Propositional Logic in considerable length, the present, second volume can stand independently on the basis of its inclusion of an adequate review of Propositional Logic systems.

Some logicians prefer to introduce Logic as a subject for study and for instruction by embarking directly on the examination of predicate logic. They then find themselves compelled to pull back and examine the fundamentals of Propositional Logic (also called sentential or statement logic.) Although such commitment to first-order or predicate logic is well motivated, it is also instructive to note the inevitability of attending to the unextended Propositional Logic itself. In the present text, rather than arrest our progress through the study of predicate logic in order to cover the basics of Propositional Logic, we begin with a brief review of the latter before we can commit our undivided attention to the former. This makes it possible for the present volume to be used independently of the first. Anyone interested in a thorough investigation of Propositional Logic on its own right could repair to the first volume, where reasons are offered as to why that project warrants considerable devotion of effort. On the other hand, if the motivation is to study the expressively more adequate predicate logic, then the present volume can be used by itself.

The formal idiom of predicate logic – also called First-Order Logic and Quantification Theory – is more expressive than that of Propositional Logic; this is accomplished by the addition of symbolic resources and the regulatory stipulation of grammatical arrangements for the management of those resources. Formalizations or translations from the language of Mathematics or from fragments of a natural language like English into some notational variant of predicate logic respect logical properties like validity, consistency or tautologousness, in cases in which the resources of Propositional Logic miss the mark. The progenitor of modern logic, Gottlob Frege, and others who followed, took the formalism of predicate logic to be fundamental. Its expressive power is the key to understanding such ardent enthusiasm. Moreover, a view took hold, by which attempted ascent beyond predicate logic was thought ill-advised. Restrictions of the truth-functional logic – which is the propositional one – by means of adding more truth values in constructing unorthodox many-valued logics have been assailed vociferously but modal logics too – which extend predicate as the latter extends Propositional Logic – have also come under vituperative attacks. Such skirmishes are of interest to the history of logic and will not concern us here. While the language of Mathematics may not need deployment of modal formalisms – since mathematical truths are not dependent on dynamic contexts like those of time or physical constraints, and cannot be impossible or actual or contingent in their logical status – this is not the case when it comes to ordinary linguistic claims, theories and arguments. The logic of language requires ascent to modal logics – even if we bypass the controversial pluralist view that discovers a variety of unorthodox logics at work in natural languages. When it comes to philosophic, or generally complicated theoretical, problems, we cannot even begin to formalize or translate their language adequately by relying on the resources only of predicate logic. The student of Logic cannot dispense with future peregrinations into non-classical logics like modal logics and possibly non-standard truth-functional logics (many-valued, fuzzy, and others.) There is no room in the present text for experimentation or expansion beyond the mere basics.

Be that as it may, the study of Logic begins with the standard propositional and predicate formalisms. After all, deep results about Modal Logic show it to be the fragment of predicate logic that remains invariant in its modeling under certain homomorphic mappings. The student of Modal Logic needs to be versed in predicate logic notation which she will encounter for sure in the metalinguistic accounts that accompany the formalism. Moreover, modal logics extend standard propositional and predicate logics – as, indeed, predicate logic is an extension of Propositional Logic. Intriguing philosophical challenges emerge especially in the case of Modal Predicate extensions but such prospects should not prejudice the study of Modal Logic itself. When it comes to expressive power, modal logics actually outrun predicate logic – as it ought to be expected – and attain the expressive prowess of higher-order systems. Specifically, we may note that, when it comes to such higher-order extensional systems, like second-order logic, we also have to omit them in the present text. A second-order formalism – in contrast to the predicate logic we examine, also known as first-order logic – includes formal resources that allow translation of predicate of relations, relations on relations, and phrases that quantify over predicates. In our predicate logic world, we will not be able to translate symbolically, to perspicuous representation of the logical structure, the proposition of a sentence like "courage is a virtue" or "two presumed separate objects are, indeed, identical if and only if they have all their properties in common." The loss in expressive power is, arguably, compensated by the better metalogical behavior of first-order logic (with second-order logic lacking a metalogical characteristic known as completeness relative to the logic's semantics.) At any rate, logical investigations must first proceed through predicate logic before future endeavors into higher-order logics can be undertaken.

The formal study of Logic continues to be neglected within the ivory precincts of academia. Few doctoral students have ever studied Logic. Prospective Law School students have an advantage in this regard since the test required for admission is a logic test – with over half the problems in the test falling under inductive reasoning. Yet, given its ancient progeny in the work of Aristotle, its unlikely progress in medieval times and the revolutionizing afforded by application of apparatus that originated in Mathematics, modern logic is an indispensable foundational propaedeutic. Because it requires extended and focused attention, and unstinted commitments of time, the task of studying logic is not a simple matter and the subject remains esoteric to most. In spite of the pledge of higher education to cultivate critical thinking skills as a learning outcome, the technical investigation of reasoning is postponed unto eternity. Analysis of the logical structure of claims, theories and presented proofs is ubiquitous in every field and discipline; its imperatives are coextensive with reasoning activities. The gap in education is damnable. As an example, being able to explain why the King, in *Alice in Wonderland*, is wrong, not merely droll or witty, when he praises Alice for her keen eyesight that allows her to see "nobody" requires some familiarity with logical-grammatical concepts and tools. The grammar of the sentence may actually be misleading with respect to its logical grammar. This is no laughing matter because the history of thought is marked with the assertions, presumed problems and puzzles, extravagant theories and critical riddles, which may be deflated when their logical grammar is scrutinized, dissected and exposed to learned inspection. Even if one does not subscribe to the provocative view of the Analytical School of Philosophy, according to which traditional philosophical problems are inevitably bound to stem from errors regarding logic or linguistic usage, still one needs to apply logical analysis to any putative theoretical challenge. The task is not confined to philosophy or to the systematic study of language, although it is paramount and unavoidable in those fields.

Given its expressive power, predicate logic is deemed as minimally adequate for formalization of such fundamental languages as that of Mathematics and for translations of the meanings of English (or natural-language) sentences. A rigorous investigation of formal logic often begins with predicate logic. Laying foundations in this area is key to a technical understanding of deductive reasoning and to honing rigorous critical thinking and problem-solving skills.

Notable (some of them unusual) features that are covered in the present volume include the following:

- To make possible use of this volume by itself, we have included an extensive Lexicon of deductive-logic terms and an overview of the basic unextended Propositional Logic.
- The overview of Propositional Logic includes positive semantic trees, in addition to the negative semantic tree method.
- Prenex forms and conversion to equivalent prenex forms are presented and utilized in subsequent sections.
- Relational (ultimately polyadic) predicate symbols, function symbols and identity are made available.
- Decision problem and the Löwenheim Result are discussed briefly.
- Proof-theoretic methods are presented analytically and extensive justifications are offered for the required restrictions on the deduction rules.
- The semantics of predicate logic modeling are presented in analytical detail along with inquiries into the logical-philosophical significance of predicate logic.
- Translation from English into the predicate logic idiom (formalization, symbolization) is examined thoroughly, accompanied by motivating linguistic observations and thorough scrutiny of available options; aspects of this inquiry include translations under restricted and unrestricted domains, translations of compacted predicates, rendering of non-standardly quantified phrases, translations of numerical statements, definite descriptions and regimentation, and guidance on how to render existential presuppositions.
- Disambiguation is imposed on translations and an extensive list of examples is presented.
- Translations of idiomatic linguistic expressions are studied.
- Semantic tree decision procedures (for finite domains) – including negative and positive semantic tree systems – are constructed and applied.
- Appendices on Set Theory, Mathematical Induction and Dialogical Logic are presented.

As a means toward teasing out theoretical subtleties and negotiating formal and philosophic challenges, this work uses detailed examples and exercises; because of this feature, the text can be used to study formal logic in a rigorous fashion. The examples and exercises – which are not solved in a subsequent appendix but often contain suggestions – serve the primary purpose of facilitating learning, which ought to be the case in the first place. The task of writing a textbook that balances reasonably complete and rigorous exploration with facilitating teachability is formidable. While I have not produced a textbook in the proper sense in this case, it might be pedagogically stimulating to attempt use of an analytical primer, like the present one, for teaching. In my experience, the counterintuitive features of deductive logic, combined with the dreaded affinity of modern logic to mathematics, gives rise to an interesting phenomenon: even if learners accomplish the task of applying mechanical and other apparatus needed for solving problems, they still remain adrift of the concepts and theory; in other words, they might be able to learn how to solve problems without having obtained a sufficient, grounding understanding of the relevant concepts – let alone the theoretical aspects of the discipline. I have wondered how this can be remedied. Excess of theory may backfire, pedagogically speaking, but the consolidation of learning requires proper laying of theoretical foundations. These considerations are in the background of the present text even though it is an analytical study of the formalisms rather than a textbook per se.

TABLE OF CONTENTS

➤ LEXICON

The subject matter of Logic compels us to begin with learning certain terms, as used in the field, before we continue. The same terms may or may not carry straightforwardly to the examination of the logic of languages – or of fragments of languages – and may or may not hold intuitive appeal. Since it is deductive logic we are talking about, the seminal and startling realization, as a threshold and stumbling block to the uninitiated, is that the properties studied by logic are determined to obtain or not by studying logical forms or structures. Empirical, descriptive, factual, informative, actual-world or any such considerations are not relevant. Interpretations can be given to our abstract objects in terms of propositions expressed by linguistic sentences, but such interpretations should never determine the properties themselves. If it turns out that empirical considerations compel re-adjustments to our formal instruments - that is a sign that our subject is simply not in formal logic.

For instance, the proposition of the compound sentence "it is Monday and it is not Monday today" is not, cannot be different, in *logical meaning* from "it is raining here and now and it is not raining here and now." Clearly, "logical meaning" is key to comprehending this adage. Both meanings – expressed by the two different sentences – are instantiations (instances, tokens) of the logical form "p and not-p" – for any individual proposition denoted by "p." The logical meaning of such a logical form – ultimately determined, let's say, by the meanings of "not" and "and" – is logical falsehood (also called invalidity and contradiction.) In any logically possible case, this form takes the truth value false. Thus, any instantiating proposition must also be false; it cannot be logically possibly true under any logically conceivable circumstances. The descriptive content of the instantiating proposition does not matter, then. If all this appears odd, let's say that logical meaning, in deductive reasoning, is a matter of true and false –

and, more generally, it is a matter of what referent value we assign to the items that matter for evaluation of the logical properties. Hence, deductive logic is a matter of form – an adage that is oft-repeated.

We study in Logic forms of arguments, forms of simple and compound propositions, and collections or sets of such logical forms. We are interested in whether the argument forms are correct – called "valid" – and whether the sets of propositional forms are consistent – so that theories instantiating them can also be consistent – and we are interested in the logical status of propositional forms (specifically, if such forms are tautologous, contradictory or contingent, as we say.) All these terms must be learned so that we may then embark on our logical investigations.

Sentence, Proposition and Formula

The word "sentence" generally connotes an item whose construction is properly achieved by specified grammatical means; such construction is not matched to considerations about meaning but it is based on rigid stipulations about how symbols are to be concatenated. Caution is needed because many texts, to avoid metaphysically loaded terms like "proposition," may use "sentence" to refer to the meaning of a declaratory sentence. In that case, semantic formal systems – in which true and false are in the base of the system – may still be said to involve sentences. If, on the other hand, we stick strictly to reserving the term "sentence" for grammatically correct symbolic-resource concatenation, we must use some other term for the meaning which well-constructed sentences express: the term "proposition" can be used for that purpose (although the term "statement" is also found in texts.) Taking all this into consideration: sentences carry meanings (propositions) but the sentences themselves are properly assembled constructs made of symbolic resources (better, tokens or instances of the available symbolic resources); sentences are correctly put together by applying grammatical or syntactical rules, which are specified for each language. This view seems to work for natural language too (for instance, for English.) In the case of studying formal logic, the term "sentence" should be used when we construct proof-theoretic systems (like sequent systems, deduction system, axiomatic or other such proof systems): there is no talk of meaning there (but, let's also point out, meaning is a matter of truth value – true and false – in deductive reasoning); instead, in such formal languages, sentences and building blocks of proof-lines are put together and carried out by means of applying specified strict rules. This is like playing a game for moving pieces but without thinking of what our pieces may represent; it would actually be a misunderstanding and a diversion to try to match pieces with objects about which we can tell stories; it would also be an error to try to make sense of the rules we use for manipulation of symbols by looking into some narrative or some independently existing realm of things; all that happens is a game with its own arbitrary but strict rules for moving around resources and moving ahead in formal procedures. In a certain respect, this is always true if we have a formal language; but in the semantic approach to system- or language-construction, we have objects to talk about, narratives to construct, and justifications in terms of those references – thus, we build models! Of course, we should not overestimate what those objects and narratives do: if they dictate how our system works, then ours is NOT a formal system! We should rather think of our modeling as having a metaphysics (a view of the world and its kinds of things) which is internal to our system and not empirically dictated from outside.

Thus, the difference comes down to this: in the syntactical-grammatical side, the rules for manipulation dictate what the objects mean – that's it and nothing else – but on the semantic side, our objects are definable in the system over truth conditions (which means that we have objects and we can narrate about our objects in a model-building fashion.) As an example: let us consider the meaning of the symbolic resource that we would match with the logic-word "not" of language (but, caution, our symbolic resources stands independently of the linguistic "not" and the matching is external and may or may not succeed): in a syntactically built proof-theoretic system, our symbol takes its meaning from the rules that regulate when it can be introduced and/or eliminated as we move around in a proof process; this is done according to some schematic recipe whose instantiations constitute application of a rule in our game. There is a view that this actually gets it right when we move outside and look into how natural language – like English – works: we understand "not" to be defined by the conditions that allow accepting justifiable use of "not" (let's say, when the sentence that is to be negated has led to a contradiction.) The semantic view, on the other hand, speaks of objects and building justifying models: it might sound odd that the fundamental objects are true and false (truth values) for meanings to refer to; also, objects we can talk about are the meanings – referents of names we might use (in predicate logic models.) Basically, meaning is understood as reference – a view that is foundational to the constitution of modern extensional logics and which, precisely because of its foundational role, attracted pronounced philosophic attention by someone like Gottlob Frege – one of the founders of modern logic.

In a well-behaved logic, the two sides – syntactic and semantic – should match each other or harmonize: what is derivable by proof-theoretic procedures should be exactly what we can model on the semantic side; and vice versa. This weighty subject is studied by Metalogic, which space considerations prevent us from engaging in the present text. The story about this desirable harmony is straightforward in the case of Propositional Logic – they do harmonize, provably – but when we ascend beyond monadic predicate logic – to relational predicate logic – matters get more complicated.

Another important feature of formal systems is decidability: whether the system can yield the targeted results within a finite number of steps that terminate according to structural rules; the results that matter concern logical properties like validity of an argument form (semantically speaking), or derivability of a thesis or theorem (syntactically or proof-theoretically speaking).

Finally, we may introduce the term formula. The flavor of this term is decidedly grammatical-syntactical. We can agree to use this term in both semantically and proof-theoretically constructed languages. A formula is a grammatically correct arrangement of symbols: better, we should then call the assemblage a well-formed (grammatically correct) formula. If we do this, we may reserve the term formula as such for any arrangement of apparent symbols – even if those symbols are not included in our symbolic language and/or are not put together in grammatically correct fashion. In that case it is meaningful to say that "so-and-so is a formula but it is not a well-formed formula." Another option is to insist that any formula – even one that is not well-formed – has symbols from our given symbolic language: then, well-formed formulas are those that are grammatically arranged correctly, in accordance with the specified, given rules of our symbolic notational idiom. In general, a well-formed formula is what we recognize – what we can scan, so to speak – and all other formulas do not register at all.

Formal Languages and Natural Languages

By now, it must have come across that the relationship between formal logic and the logic we might expect to find embedded in a natural language like English is a somewhat complex matter. We don't study formal logic to figure out the logic of a language. A surprising view that has prevailed is that it is indeed a matter of discovery what logic or logics are at play in a language (we are not deciphering some aloof entity called Reason); but, on the other hand, we do not embark on this discovery by putting together formal systems. Let us assume that the logic of a language – not to insist that there is only one such logic embedded in a language, but some logic – is called X. The statement "language L has logic X" can possibly be true and can possibly be false (of course, not true and false together but possibly the one and possibly the other.) On the other hand, nothing about a formal language is possibly true and possibly false. A formal system of logic or a formal language is well-regulated in a rigid fashion and yields its internal results with rigorous inevitability; defects in such a system are formal – like those studied by Metalogic and, as mentioned above, may consist in decidability failures or in failures to harmonize with the semantic side of the logic. Thus, it is a contingent matter whether some logic X is at play in a language L but there is nothing contingent about the results we derive when we study the inner workings of a formal system. We may think of what is happening here along the lines of how a geometrical system like Euclidean Geometry relates to the "real" geometry of our universe; it may or may not be the physical geometry of our universe (indeed, Relativistic Physics had to draw on a different geometry and, so, if Relativistic Physics is the physics of our physical universe, Euclidean Geometry is not our physical geometry – but, note, there is and there can be NO internal damage or falsification to the geometry itself!) Similarly, it is not shocking to discover, after all, that the Propositional Logic we might have studied doesn't get it "right" when it comes to the logical behavior of all uses of "not" in the language. But, to the extent, that our formalism does get it right, we have a powerful, indispensable tool for logical analysis of our linguistically based claims and theories.

Logical Forms and Logical Constants

It is often said that deductive logic is a matter of logical form; this also implies, properly, that deductive reasoning does not depend on content (what is being talked about). This is consistent with the other declamation often made – that deductive reasoning does not depend on empirical, factual, or descriptive matters. This is an initially surprising and counterintuitive claim and it serves as stumbling block to the study of logic, which turns out to be one of the difficult subjects. To understand what is happening, we can think of a sentence of English like "it is raining and it is not raining." We don't care about the linguistic grammar but about the logical grammar and this latter depends on the meanings of the logic-words in the sentence. Such words are the ones whose meaning or definition is indeed a matter of how they interact with true and false (the truth values.) A word like "not" cannot be matched with some item – like "dog" as a token of the way we write the word referring to a dog could be matched with an image of a dog. We could stipulate picture-like symbols, instead of "dog," toward a hieroglyphic-like language but when it comes to words like "not" we would find ourselves at a loss as to how to be "realistic" in our experiential referents for the word. Indeed, the meaning of "not" is: it turns true to false and false to true. Hence, this is a logic-word. The same is the case with "and" – and with no other word in the above sentence. If we underline those words, and only those words, we have: "it is raining <u>and</u> it is <u>not</u> raining." These are the fixed words but nothing else is; the rest can be replaced by lines (placeholders, variables): "__ and not__." Any single sentence can be put into each of the lines. Because we notice that it is the same sentence that is negated, we use the same variable or placeholder or line-type. What we have now is an unofficial logical form – it is not written in some specific formal language but the point is that only the logic-words are fixed while the other parts can have *any* sentences plugged in. Thus, any sentence plugged into the line will yield one of an infinite number of instances of this logical form. This form has a pronounced logical feature: it is a logical contradiction which means that it is false regardless of whether its component sentence is true or false. Notice that the content does not matter; any sentence can be plugged in. Always you get a logical contradiction in the sense given above. The only fixed parts that matter as contributors to the meaning are the logic-words, "not" and "and." Thus, we don't have dependence, indeed, on content or on empirically verifiable or on factual matters. (Of course, the meanings of the logic-words are given: you could quibble that this too stems from experience but let's say in response to this, and let's specify, that the meanings of such logic-words are presupposed and are not considered as discoverable. The basic language – the logic-words definitions – are there to begin with; they come, fixed, for the game to be there in the first place.)

As a term, "constant" does not appear much in the bibliography anymore. When we deal with a semantically constructed system of Propositional Logic, we make available symbols for what we call connectives. These interpret algebraic functions of a special algebra that has only two values, from the set {1, 0} and so that true interprets one number and false interprets the other. In a syntactically made proof-theoretic system, we should use the term operator instead. The connective – or operators – are retained when first-order or predicate logic extensions are built. Predicate logic also makes available symbols for logical-predicate symbols, for names and variables and possibly also includes the identity symbol. The predicate symbols are often called non-logical or predicate constants. Here is a deeper, more insightful way of making sense of why the predicate symbols are considered non-logical constants but the connectives are logical constants (if we wish to call them so, although this is now unusual.) We can think of logical concepts as being those that remain invariable across transformative shifts of the right kind. Of course, it is difficult at first to discern what we mean by "the right kind." Let us think of a logical predicate like "is a student" – presented informally since we are not making yet any symbols available to us and we are talking about this in English. The logical meaning or value of "is student" has to be a referent in accordance with the extensional character of our formalism. By a clever trick, we declare the referent of "is a student" to be the set that has the students as its members – and no has nothing else as a member. Of course, there might be a price to pay for mobilizing the machinery of sets but that is outside our present scope of inquiry. Now, we can question which students we include: present, past and future? Only those in a specific class? And so on... We can think of an indefinite number of available universes or domains. We can have a domain in which no one is a student – hence, the value of "is a student" is the empty set. We can think of another domain in which everyone is a student: there, the value of "is a student" is the universal set. And we have an open-ended number of options. Thus, the meaning of "is a student" – which is its referent or value – is variable. This is not the case with our negation function, for instance. Thus, negation is a logical concept but "is a student" is not. This also means that "everyone is a student" does not express a logical truth or a logical falsehood (it can be true in some models but false in other models.) This is exactly what we expect. On the other hand, "everyone is a student and no one is a student" cannot be true in any model – it is a logical contradiction, which is, again, what we would expect. What are the logical notions that render this last proposition a logical falsehood? Those, and only those, are the logical constants: the meaning of "not" and also the meanings of "everyone" and "someone". Thus, although "is a student" is not a logical notion but a non-logical predicate constant, "everyone" and "all" are logical notions: a pleasant approach to appreciate this – provided we confine ourselves to finite domains only – is by thinking of "all" as an "and" and of "some" as an inclusive-sense "either or." "Everyone Fs" can be thought of as "a_1 Fs and a_2 Fs and ... and a_n Fs" for a domain of n objects with "a_i" as the name of such an object. It is an interesting exercise to ask ourselves if the names – also called constants – are logical notions or not. They shouldn't be!

Logical Consequence, Argument, and Implicative Propositions

We can say that a logic is characterized by its relation of logical consequence. We can think of this as the collection of all the argument forms that are valid in the logical system we are discussing. An argument form represents a relation – between premise-forms and the conclusion-form. Validity of an argument form – as a foundational concept in the study of logic – is what we may intuitively grasp as "correctness": a sanctioning that the conclusion is true necessarily (as a matter of logical necessity) if the premises are all true. At a minimum, it should be logically impossible to have all true premises and a false conclusion for any instance of a valid argument form. Thus, even one instance of a given argument form, which has all true premises and false conclusion, refutes the claim that our given argument form is valid. Such an instance is called a Counterexample – but this term has other, formally precise definitions too as we will see. We could check for counterexamples to given argument forms until we might find a counterexample but failure to find a counterexample does not establish that the argument form is valid – it could be our heuristic limitations that prevent us from hitting on a counterexample. It might seem that recourse to empirical or factual matters is significant, after all, since our search for counterexamples proceeds through instances of the argument form – and those instances may well include statements about empirical matters, so that the statements are given their truth values on the basis of actual information. Nevertheless, this is a misunderstanding. Consider that we can also produce a legitimate, and equally damning, counterexample by making up alternative, non-actual, states of affairs and by discovering, within such contexts, an instance of the given argument form with all true premises and a false conclusion. Thus, it is not actuality but the invalidation – all true premises and yet false conclusion – that matters and the search for counterexamples is something of an ancillary device that forces reflection on the logical possibility of having an invalidating case.

Generally, logical possibility should be thought of as conferring equal rights, so to speak, to every consistent narrative – no matter how outlandish such an alternative world, so depicted, is. In practice and outside of introductory Critical Thinking studies, we do not depend on a search for counterexamples. Formal systems that are characterized by completeness – relative to the semantics of the logic – are able to produce counterexamples for all cases, and only the cases, of invalid argument forms: here, a counterexample is defined as the set of values for the components of the statements, for which all premises are true and the conclusion is false. In the case of Propositional Logic, the only values we are dealing with are the truth values (true – false)) for the atomic (ultimate individual) propositional variable letters of the given formulas.

Another approach to the characterization of a logical system is by considering the set of its logical truths or tautologies (addressing the task semantically.) Caution is needed about something, though. There are non-classical logics for which the so-called Deduction Theorem does not apply: this means that we can have logical consequence (logical impossibility of all premises X1, ..., Xn being true while the conclusion Y is true) without having the characteristic implicative proposition expressed by "if X1 and ... and Xn, then Y" being a logical truth of the system. In that case, our characterization of the system in terms of tautologies would be wrong; the deeper characterization, it turns out, is based on logical consequence. Indeed, there are unorthodox logical systems that have no tautologies. This sounds odd but consider that some third value (besides true and false) is added as another way of "failing" besides false, which is also a failing or non-satisfying value. Depending also on how the connectives are defined, we can have always the possibility of assignment of truth values to the atomic parts so that the whole formula receives a non-satisfying value: hence, there are no tautologies whatsoever. And yet, such logical systems do have a characterizing logical consequence relation: there are argument forms for which there is no assignment of truth values to the atomic parts, for which all the premises are satisfying while the conclusion is non-satisfying. (We also note that the distinction between satisfying, or better called designated, and non-designated values is broader than the distinction between our classical logic's true and false.)

There is no deeper reason why we can only have one conclusion in argument forms. There are formal systems that can manage more than one conclusion. We ought to keep in mind that the premises in the argument forms are conjuncts and the conclusions are inclusive disjuncts: in other words, think of the premises as being joined by conjunction (linguistically by "and") and think of the conclusions as joined by inclusive disjunction (linguistically, "either-or" which means "or/and" – not the other meaning of "either-or" which means "one or the other but not both.")

The characteristic logical property, of argument forms, which we will be examining is called validity: a valid argument form in the two-valued standard Propositional Logic is logically-necessarily not-invalid: thus, it is not logically possible for any of its instances to have all true premises and false conclusion. In the familiar decision mechanisms, often the search is for a counterexample: values for the atomic parts, such that the premises are all true and the conclusion is false for that value-assignment. Discovery of such case – which is a counterexample – determines that the given argument form is invalid. Failure to discover this, in a mechanical and complete decision procedure, establishes that the given argument form is not invalid – therefore, it is valid. Obviously, we are taking advantage of the polarity between valid and invalid – reflecting that between true and false. There are logics in which we can actually define degrees of validity – a philosophically controversial approach – but such logics are, by definition, non-classical and they fall outside our present purview.

Consistency

Another logical property we explore in the study of logic is consistency: this is a logical property of collections of propositional formulas. Thus, if an argument form is given for scrutiny, we cannot speak of consistency (although we can speak of consistency of its premises or of consistency between premises and conclusion or between premises and negated conclusion - but the argument form is a relationship between the premises and the conclusion; thus, in argument forms, and only in argument forms, there is a distinguished formula which is presented as the putative conclusion.) A theory can be thought of as a collection of propositions; hence, the question of consistency arises with respect to a theory and it is a crucial subject. An inconsistent theory ought to be rejected - if it is not modified so that the new theory is not inconsistent. As with validity and invalidity, an inconsistent theory cannot possibly be consistent and vice versa. The definition of consistency of a collection of propositions is: it is logically possible for them to be all true together. If we were to give an interpretation of the given propositional formulas by means of actual-world sentences, we could find out that they are not all true: but this does not prove that we have an inconsistent collection of propositional formulas! As before, the actual empirical state of affairs has no privileged status: it is one logically possible world and nothing more. We could have other logically possible worlds - or narratives that exhaust descriptions in such a logically possible world - in which the interpreting propositions are all true; in that case, we have established that our given collection is indeed consistent. It is not easy in practice to determine whether a collection of propositions - a theory, we can say - is consistent or not. If our theory is inconsistent, which is something we might not realize at all, then we have in our hands a logically pathological case! In the ancient dialogues written by Plato, his character Socrates reduces offered theories to absurdity: he shows how, for some hypothetical scenario that gives us interpretations for the theory, we can actually prove validly a contradiction from the theory. A contradiction has a logical form that cannot possibly be true. The most familiar case of a logical contradiction form is "p and not-p" for any proposition denoted by "p." It is indeed the case that, if the theory is inconsistent, and only if it is inconsistent, a contradiction is provable from it. Decision procedures to check for consistency are available, which directly determine if the given propositional formulas can possibly be true together. If there is no logical possibility of this, the theory or collection of propositions is inconsistent. Terms like "logical possibility" are actually tricky but the decision procedures of formal logic have precise and provable correct ways of checking for such matters.

Now we can also bring the concepts of validity (of argument forms) and consistency (of collections of propositional forms) together. To see the connection requires some reflection – and, certainly, comfortable familiarity with the concepts. It turns out that an argument form is valid if and only if the collection of its premises and the negation of its conclusion is inconsistent (in other words, it is not logically possible for its premises and the negation of its conclusion to be true together – which means, of course, that it is logically impossible for all its premises to be true and its conclusion to be false.) When we say "at the same time" or "while" or any other such seemingly temporal tropes of speaking, we mean really value-assignments (also called cases, options, and even logical possibilities.) There is nothing dependent on temporal accidents or actual events in all this. The game is to assign values to the components of our formulas and examine, for all cases, what values we get for the compound. We are able to do this in Propositional Logic and – up to a point – for predicate logic. This characteristic of the standard logics is called – for obvious reasons – Compositionality of Meaning. We don't have this as a characteristic of linguistic compounding of meanings – so that the meaning of the whole can be correctly and fully determined from the given or specified meanings of the components. And, of course, meanings are truth values in Propositional Logic - and we will expand this conceptualization for predicate logic when we get to it.

Logical Status of a Proposition

We are also interested in examining what we may call the logical status of a given proposition – better, of its logical form. There are three possibilities – and, hence, three cases with their terms to learn. If the propositional form is true for all logically possible cases (specifications of its component values), then we have a tautology or logical truth (speaking semantically.) If the propositional form is false for every logically possible assignment of values (valuation) of its component parts, then the form is a contradiction or logical falsehood. The last case is one in which the form is true for some and false for some other valuations: this is the case of what we call a logical contingency or a logically indefinite or logically indeterminate propositional form (also called possibly-true-and-possibly-false.) It does not matter how many case of valuations yield true or false – what matters is that we have both logical possibilities available. All informative, descriptive, factual sentences of a language should indeed have logically contingent forms. Tautologies and contradictions are – as we may say – logical truths and logical falsehoods, respectively, and they are the logical forms of propositions that cannot convey any empirical information. Notice that, although contingent propositions have contents that are not the business of logic, so to speak, the determination that a propositional form is contingent is a matter of logical investigation. A disputation over a logical truth or a logical contradiction – which can happen, for sure – ought to count as a pseudo-disagreement, not as a genuine disagreement. This is not to say that the disputants are aware of this; it is a fair question, how many acrimonious confrontations, even violent cases of opposition, are predicated on pseudo-disagreements that stem from inability to detect characteristic logical status of the propositions used – and similarly with respect to systematic inability to detect if theories are inconsistent and arguments are invalid.

It may appear that logical truths and logical falsehood ought to be knowable independent of experience – and, as such, they are to be called *a priori*. This is a normative subject – about what is properly the case, not necessarily about what actually happens. Individuals may not know certain things that are knowable a priori. Traditionally, logical truth and logical falsehoods have been associated with a priori knowledge. Of course, in the case of logical falsehoods, they cannot be known (only truths can be known); we should say that their negations are knowable a priori. Nevertheless, it turns out matters of knowledge (epistemic matters) pertaining to the a priori are separate from logical characterizations. This makes some sense if one were to insist on the apparent separation of the realm of knowledge from that of logical subjects. It would take, however, some examples of logical truths that are not knowable a priori. Such a case is furnished by an initially surprising result in modal logic – and, as such, it lies outside our ken. There is an old proof – which met significant resistance – by which if two names refer to the same thing, then it is a matter of logical necessity that they do so! There is a way of accounting for this which makes the oddity disappear. At any rate, this would be the kind of example we are looking for: it is discoverable by experience whether, for instance Morning Star and Evening Star refer to the same celestial object (also known as Venus); so, here is a logical truth (that they necessarily co-refer) which is not knowable a priori.

Another term we come across in the study of logical properties is analyticity. A proposition is analytic if and only if its truth value (true or false) can be determined simply on the basis of the meanings of the words in the conveying sentence. For example, "a triangle has three angles" is analytic and analytically true whereas "a triangle has four angles" is analytic too and analytically false. We can make the following observation: insofar as the meanings in question (like "triangle," "three," and "angle" in the preceding example) are non-logical words, we take such analytic propositions not to be the business of logic: it is not a matter of the definitions of the logical words, it is not a matter of logical form, and clearly the form of the proposition in the example is not a tautology, or a contradiction. When the meanings of the logic-words in the sentence (like "not" and "and," for instance) suffice for determining the truth value of the whole, then we have our familiar tautologies and contradictions. Thus, we may say that the concept of analyticity is broader than that of logical status. Analyticity has been subjected to philosophic challenges but it is safe, given our present scope of inquiry, to assume that the narrower slice of analyticity we investigate (when, for instance, we decide logical status) is free of any challenges or pressing problems.

Logical Relations

Two propositions are mutually equivalent if and only if they imply each other; this means that they have the same logical meaning (truth value) for all logically possible assignments of truth values to their ultimate atomic or individual propositional components.

A proposition is said to imply another proposition if and only if it is logically impossible for the first proposition to be true and the second to be false for the same assignment of truth values to their ultimate atomic propositional components. If, of two propositions, neither one implies the other, the propositions are said to be logically independent of each other.

0. REVIEW OF PROPOSITIONAL LOGIC

The most basic logic is Propositional Logic. The study of the logic of language, or of some fragment of a so-called natural language, is a worthy pursuit on its own right but the student of formal logic detects that the subject matter is some artificial system or symbolic language. To be precise, what we study are notational variants of systems of Propositional Logic. In that sense, every text includes its own variant – and, indeed, more than one such variants – and it should be deemed coincidental if such systems overlap in their symbolic resources and grammatical arrangements. Unless, of course, an author identifies by name and reproduces some formal system that has already been presented.

Propositional Logic itself can no more be constructed as a formal system than "the triangle" can be presented on a piece of paper: instead, we deal with tokens that exemplify or instantiate the conceptually constructed abstract object we call "triangle." These are subtle points, perhaps, but completeness dictates that we make such matters explicit. Whether our formal idioms correspond to the logic or logics of language – or of fragments of language – is actually an issue that is external to the formal study of logic itself. This sounds odd at first. We may compare the study of the system bequeathed to us from hoary antiquity and known as Euclidean Geometry. Whether the geometry of our physical universe is indeed Euclidean or not – it turns out that it is not insofar as Relativistic Physics is accepted as the physics of our universe – is a separate matter from the study of the geometry itself. If it turns out that our physical geometry is not Euclidean, after all, that does not redound against the geometrical system in any way whatsoever; the empirical discovery only falsifies the statement "the geometry of our physical universe is Euclidean," and this sentence was always to be assessed outside of the statements of the geometry, so to speak.

The separation of logical formalism from natural language is, similarly, a foundational caveat. When a logical language like that of the basic or standard Propositional Logic is attacked as missing the mark when it comes to the analysis of logic-words of the natural language (notoriously, in the case of the "if-then" logic-word), we could charge that the critic has misunderstood the subject. As in the case of geometry, our formal system of Propositional Logic is studied from within, so to speak, and, indeed, if this were not the case, we could not claim to be studying a *formal* system. To the alert and informed student of the subject, this result is eminently satisfactory because it comports with the expectation that our deductive formal system - as the case is generally with deductive reasoning - is immune to empirical, factual, descriptive or any related interventions, corrections, and objections.

In fact, the familiar Propositional Logic is not the minimal logic of propositions. Intuitionistic Logic can be presented as a sub-logic of the standard Propositional Logic in proof-theoretic systems and there is a formal logic, known, in its uninterpreted variety, as the Lambek Calculus that is minimal with respect to the other logics. By relative minimality we mean that a logic is a sub-logic: its logical truths (if it is presented semantically) are included in (are a proper subset of) the logical truths of the standard Propositional Logic but there is at least one logical truth of the standard Propositional Logic that is not a logical truth of the sub-logic. There are other ways to define this concept but, for our current purposes, we can rest the case. It has turned out, in the context of some ambitious applications of a formalism known as Generative Grammar, that the Lambek Calculus - or, rather, an interpretation of it - comes handy and can be applied when it comes to studying the logic of language in a more detailed manner than had been thought feasible in the tradition. We cannot pursue this subject here but the moral – as above – remains that the vagaries of applying logical formalisms to the study of the logic of languages are a separate issue from the formal study of logical systems themselves.

Notwithstanding the above remarks, we can count on the useful application of formal idioms of logic to the logical analysis of structures of arguments, theories and claims presented in natural language. Then, the pressing issue is adequacy of the formal logic to the task of studying logical structures in various linguistic contexts. Propositional Logic has been thought rather inadequate. A deeper issue in all this has to do with the availability, within a formal system, of symbolic resources. Of course, whatever resources are made available must be regulated by the specified grammatical arrangements of the formal system and this is not a trivial matter. Because Propositional Logic lacks symbolic resources for the representation of the internal ingredients of propositions, it is straightforward that Propositional Logic is less adequate for the study of the logic of language in comparison to predicate logic which makes additional symbols available.

Due to its lack of symbolic resources, Propositional Logic is reduced to translating certain arguments that are valid in such a way that they are assessed, within the impoverished Propositional Logic, as invalid. A parallel problem of relative symbolic insufficiency arises when we contemplate extensions of propositional and predicate logic to systems of Modal Logic. There have been objections, now mostly put to rest, against modal logic. Another richer logic, Second-Order Logic (or higher-order logic of some sort of other), has also been subjected to formal criticisms. The crucial issue is that our motivated interest in enriching our symbolic resources is counter-balanced by the emergence of defects in the more complex formal systems. Discussion of such defects is controversial and is of interest to Metalogic – the study of formal languages – but it is also involved in philosophical-logical debates. Arguably, predicate logic itself presents certain metalogical defects related to decidability – arising prominently from the prospect, lacking in Propositional Logic, of infinitarian constructions. In spite of related problems that arise regarding decidability, predicate logic is regarded as unproblematic and, indeed, as the least (and some argued, the most) enriched symbolic formalism we should make available for the study of formal logic. Predicate logic confers a role, not found in Propositional Logic, to what we may call non-logical words: if we think in terms of the logic of natural language, a word like "not" is logical whereas a word, for instance, like "student" is non-logical. There is an intuitive appeal to this distinction. One might suspect that there are inherent risks to the availability of symbolic resources for what we call logical predicates (interpreting the phrase "being a student", to use our preceding example): although the enrichment is welcome, allowing for more adequate translations in preparation for determining logical status or argument validity, on the other hand, the gain depends on introducing non-logical items. Nevertheless, the contamination by symbols referring to items that interpret non-logical parts of language is managed without loss to the fundamentally logical character of the formalism. We may inquire as to what this fundamental character of logical things is. A cursory answer is that logical objects are those that remain invariable across the kinds of transformations that are available in the formalism. Thus, even though non-logical predicates can be handled variously across different "stories" we can tell, the logical concepts are not affected: truth and falsehood are to be understood as truth-in-all-stories and falsehood-in-all-stories respectively. Thus, we realize that we are indeed dealing with non-empirical and properly invariable notions. It is logical truth and logical falsehood that we are dealing with in spite of the introduction of non-logical items.

Since we will concentrate on Predicate Logic in this volume, we only allow ourselves a brief overview of Propositional Logic, taking advantage also of this opportunity to settle some basic terminology in Logic. The fact that predicate systems extend formal constructions in Propositional Logic compels us also to lay out the formal resources and grammars of the erstwhile propositional languages.

An important distinction we should keep in mind is between proof-theoretical and semantic approaches to the building of formal languages. Other terms for proof-theoretical or proof-theoretic are syntactic or syntactical (although it should be kept in mind that any formal system is organically dependent on syntactically mandated manipulations of symbolic items); deduction-systems; proof-systems or proof theories; sequent-systems. Terms of the art we should associate with the proof-theoretical approach are derivation, inference, deduction. In such systems, we speak of rules and the philosophic support for insisting that Logic should be studied along such lines rests on a claim that logical notions are definable through rules that enforce the proper kind of manipulation of symbols. On the other hand, the semantic or semanticist approach takes logical notions to be definable in terms of truth conditions which are characterized by systematic dependencies of observations on assignments of values to components of propositional formulas. When we are given that the logical connectives are defined over truth values, for instance, we realize immediately that we are dealing with a semantically constructed formalism. Semantics for a logic compels us to speak of validity of arguments – rather than of derivability – and of logical truths – rather than of theses or theorems. It is in the semantic approach that we build models. So familiar a device as the truth table is a modeling of sorts. When presented with a derivation- or proof-system, it is imperative to establish – in the field known as Metalogic – that the system permits derivational construction of all and only those proofs that correspond to valid argument forms on the semantic side.

If we think of the semantic side as laying out what *should* be provable, the syntactic side supplies us with the formal construction of what we *can* derive. What should be the case and what is the case ought to harmonize, coincide, and parallel each other. If we can prove everything that should be provable in this sense, we have a complete derivation system. If we fail to derive exactly what should not be provable, we have a sound derivation system.

We can now proceed with definitions of terms and formal grammars in the context of this brief overview.

A proposition (alternatively, statement) is defined as the meaning of a declarative sentence (a sentence that is meaningful and, as such, is rightly assertable in discourse if, and only if, it is true.) From a strictly formal standpoint - disregarding the relationship between symbolic logic and language or the linguistic orientation toward the empirical world - a proposition is a semantic interpretation of a sentence (with the sentence understood as some arrangement of symbols, which is correct according to a specified grammar.) There are simple (individual, single, atomic) propositions and there are compound (complex) propositions that are constructed from conjoining propositions. In language, the conjoining is carried out by means of logic-words. A logic-word like "not" generates a compound when applied to a simple proposition even though, grammatically, in a language like English we cannot think of it as a connecting particle. The reason is that the definition of "not" is based on truth conditions (defined, indeed, as "turning true to false and turning false to true.") This shows us that this is a logic-word since its meaning is dependent on truth-conditions. Tracking the logical grammar of a proposition requires unpacking or decomposing or parsing the given proposition so that "not" is separated from what remains. This is an acute reminder, and illustration, that logical grammar and linguistic grammar do not coincide; indeed, linguistic grammar can be misleading with respect to the logical grammar as our preceding example shows (since a simple sentence with "not" in it is compound from a logical-grammatical standpoint.) The logical meaning of "The earth is not flat" depends on the logical meaning (true or false) of the proposition "The earth is flat." The decomposition of a proposition is not a side benefit but it is actually deeply characteristic of the logic: the meaning of the whole (truth-falsehood of the compound) depends determinately on the meanings (truth-falsehood) of the parts or components. Clearly, this is not characteristic of the linguistic-grammatical structure: we cannot expect reliably to discern the meaning of the whole on the basis of some grammatical decomposition of a given sentence.

We construct a notational idiom of Propositional Logic, which we designate as PROP. All the connectives of Propositional Logic are truth-functional. This is the case of non-standard (also called alternative and unorthodox) logics that define their connectives over more than two truth values. When we speak of connectives, it should be clear that we are thinking in semantic terms. Connectives – also called logical constants in older texts – are defined over truth values. The standard Propositional Logic makes available exactly two truth values, true and false. Although is appeases our intuitions to call those truth values by the names "true" and "false," it is an interesting challenge to try to understand why the choice of names is not important. What really matters is that one of the two truth values, the one we call true, is, as we say, designated: this truth value satisfies, as we say. In terms of our understanding of the concept of validity of argument forms, this means that the designated truth value – true – is the one that gives to formulas the property that must be preserved from the premises to the conclusion in the case of validity. This means that, in a valid argument, and only in a valid argument, the truth of the premises necessitates logically the truth of the conclusion. Another way of saying this is: in a valid argument, and only in a valid argument, it is logically impossible for all the premises to be true (to have the designated truth value) and for the conclusion to be false (to lack the designated truth value.) Two-valued or bivalent Propositional Logic has an appealing simplicity to it in that it has exactly one of two kinds of values - one designated and one anti-designated. Invalidity of an argument means that the crucial property - designated truth value, true - can possibly be possessed by all the premises and yet lacked by the conclusion. Thus, preservation of the designated truth value fails. Of course, validity is a matter of argument form and the reason valid arguments are logically necessitated to retain or preserve the designated truth value (true) from premises to conclusion is because they have a logical form that has - and confers - this characteristic.

The connective symbols are given in the grammar of PROP, presented formally below. The symbol denotes a truth function, so that for any unary connective symbolized by "*" and binary connective symbolized by "^", their functions are defined as follows:

$*(x): \{T, F\} \longrightarrow \{T, F\}$ /// $\wedge(x, y): \{T, F\} \times \{T, F\} \longrightarrow \{T, F\}$

Of course, we will not be using functional notation as shown above. These present remarks are transacted in our symbolically enriched metalanguage. What we see above is that the domain of a truth function is the set of truth values or - for n-place truth functions - the Cartesian product $\{T, F\} \times \ldots \times \{T, F\} = \{T, F\}^n$ and the range is always the set of truth values $\{T, F\}$. [See Appendix on Set Theory.] Because of this restriction to truth functions, our formal language does not have symbolic resources for translating non-truth-functional expressions of a natural language like English. Not that we should assume that our translations - even for what we are able to translate - are guaranteed to capture adequately the content of the propositions that are expressed linguistically. Our target is, rather, to capture the propositional content that is relevant to the logical structure of the proposition we translate. When it comes to non-truth-functional expressions of language, however, we are not at all able to formalize such parts. The check as to whether we are dealing with non-truth-functional expressions consists in producing at least two different cases in which the same truth-value inputs yield different truth-value outputs. To do this we may rely on picking linguistic propositions and translating them. This does not introduce any empirical or extra-logical constraint into our proceedings. (Compare, for instance, how we can informally produce a counterexample to an argument form by finding some linguistic instantiation of the argument form, in which all the premises are true and the conclusion is false. Reliance on examples from language does not undermine our claims to be studying deductive reasoning: the specific examples themselves do not matter; what matters is that we show that it is possible to have all premises true and a false conclusion.)

For example, the expression "necessarily" can be shown to be non-truth-functional: we have one true proposition expressed by the sentence "a triangle has three angles" which yields also true as output for "necessarily(a triangle has three angles)." Yet, we also have, for instance, a true proposition, expressed by "the capital of Germany is Berlin" which is not necessarily true and, thus, yields false to "necessarily(the capital of Germany is Berlin)." By necessarily true we mean the semantic characteristic of linguistic propositions whose truth value, as true, is determined trivially based on the meanings of the words in the sentence: such propositions are called analytic. We also include propositions whose meanings are determined based on the meanings of the logic-words in the sentence, in which case we are dealing with logical truths. Although there are some controversies swirling around the concepts of analyticity and logical truth, we bypass such matters here.

Since Logic can be defined as the study of certain logical properties like validity of argument forms, consistency of collections of propositional forms and logical status of propositional forms, we should offer the definitions of such concepts. It is said often that deductive reasoning is a matter of logical form or structure, and is unencumbered by any considerations that rest on empirical, factual, informative, or descriptive matters. For instance, any proposition expressed by a linguistic sentence that has the form of a conjunction of a (simple or compound) proposition and its negation is a contradiction and, as such, it is necessarily false. It is irrelevant what the content of the linguistic sentence is. It is the logical form that disposes of the issue about the logical status of the proposition (if this proposition is a contradiction, for instance, or if it is a logical truth.) It is also a matter of logical form if the logical status of the expressed proposition is logically indeterminate. Propositions that have this status are the informative, contentful propositions of language. Nevertheless, the fact that they are indeterminate, so they can perform their informative role, is itself independent of what their content is: the proposition is indeterminate because its logical form can be true and can be false in different cases – it is possibly true and possibly false in a logical sense. Although it is customary to speak of logical forms, we can trace logical properties, and characterizations of different logical systems, ultimately to the defined meanings of the connectives in a logical language.

An argument form is valid if and only if there is no logical possibility that any instance of this form can have all true premises and a false conclusion. A collection of propositional forms is consistent if and only if there is a logical possibility that all propositions are true in the same case – or there is a logical possibility that their conjunction is true in some case. We will see that propositional models that are afforded us by such methods as the familiar truth table show us perspicuously what we mean by cases (or options or logically possible states – which are to be thought of, in the case of Propositional Logic, as discrete and complete assignments of truth values to the ultimate constituents of propositional formulas, which are the atomic or individual propositional variables or variable letters.) It is an overarching dictum that logical meaning is a matter of truth value (true and false) and perusal of these basic definitions we have given ought to make this evident.

0.1 A FORMAL LANGUAGE FOR PROPOSITIONAL LOGIC (PROP): GRAMMAR

We construct first a symbolic notational idiom for Propositional Logic, which we designate as PROP. Since Predicate Logic is an extension of Propositional Logic, we will consider this PROP system as incorporated within our subsequent predicate logic language. The basic formal language, PROP, has atomic or individual variables - letters from the denumerable set {p, q, r, ..., p_i, ...}, with subscripts from the denumerable set of positive integers, as needed, for single or uncompounded propositions; we also include symbols for propositional connectives, comprising one of the four mathematically definable connectives and four of the sixteen definable connectives – {~, ·, V, ⊃, ≡}. Determination of these numbers of available connectives is based on the fact that our connectives are defined over the set of two truth values {T, F}. Indeed, the standard Propositional Logic is bivalent or two-valued. Because our definitions of the connectives are over truth values, the approach is semantic. The connective symbols interpret semantically Boolean functions which are themselves defined over the set {1, 0}. The choice of the numbers is not important; that we have exactly two numbers is important – and so is the construction that forces us to take exactly one of the values as "winning" and one as "losing". We will specify more precisely what this means subsequently. While the uninterpreted algebra does not need some intuitive – or, at least, viable – narrative to accompany it, the case with semantics is different. In semantic approaches, we are able to tell stories – so to speak – about specifically designated objects. We should notice, however, that one would miss the point if he or she were to become interested in those objects themselves. The enterprise is not metaphysical. Something else is happening. For all intents and purposes, we may think of our objects as abstract. The Boolean origins of our formal system remain in the background and are not highlighted at all when the approach is semantical. It might also seem that we are taking our bearings from language – possibly English – or that we are exploring and codifying the logic that is embedded in the language. But, as we will see, this is not what we are doing. Our construction is not actuated by empirical observations. What we are doing is more like Math than like Physics – although, to be sure, the development of systems in Physics turns into Math early enough in its vagaries.

The system of connectives in PROP is redundant in the sense that some of these connectives can be defined from other connectives in the set. Using a set like {~, ∨}, {~, ·}, or {~, ⊃}, among others, would have forestalled redundancy but it would incur a cost in the relatively more unwieldy character of the formulas we would need to write. We speak of connectives but, occasionally, we might also speak of connective symbols. There is a school of thought that sees logical enterprises as consisting exclusively in the manipulation of symbols, so that meanings too ought to be defined in terms of the resources and formal actions that involve the symbols of the system. This is a fascinating philosophic-logical subject that we do not have room to investigate here.

For translation of individual or single propositions expressed by meaningful assertable (declarative or declaratory) sentences of English we use as symbolic resources capital letters from {A, B, C, ..., A_i, ...}, with subscripts, as needed, from the positive integers. As always in Propositional Logic systems, the sets of variables are denumerable – which means that their size is that of the natural numbers: what this means is that we can match each one of our set's members with exactly one natural number ad infinitum.

Finally, we require parentheses, used only to prevent ambiguity, from {(,)}; well-formed formulas necessarily have equal numbers of left and right parentheses. Insofar as the connectives of the system are defined over the set of truth values {T, F}, our system is characterized semantically and it is bivalent or two-valued. In proof-theoretic systems – like the natural deduction systems in subsequent sections – we regard the symbolic resources as being mere squiggles, so to speak, and their manipulation is governed by systematic regulatory recipes – which are rule-schemes. We should be using different symbols, since we are dealing with distinct systems, but, for the sake of convenience, we allow separate formal systems to share symbols - regarded officially as a mere accident of homonymy.

GRAMMAR(PROP): $p_i/A_j/\{\sim, \cdot, \vee, \supset, \equiv\}/\{(,)\}$

We talk about our formal system in a symbolically enriched metalanguage which is a fragment of English with added metalinguistic variable symbols from $\{\varphi, \psi, \ldots, \varphi_i, \ldots\}$ to denote not necessarily individual propositional compounds. Our Metalanguage may also be enhanced with other symbols and also with symbols from our Object Language PROP, as specified.

Our formal grammar is now presented precisely. Anything that violates this grammar cannot scan – it is to be regarded as nonsense – from the standpoint of our formal language PROP.

- $\ulcorner p \urcorner$ is well-formed

  [There is an interesting reason for using the special metalinguistic symbols around the propositional variable. They are called <u>Quine corners</u>. This is used only when the symbols are from our Object Language. We are making tokens of such symbols available within the Metalanguage in this fashion. On the other hand, you will observe that for variables φ and ψ, which stand for any well-formed formulas, not necessarily single but possibly compound too, we use standard quotation marks when we mention them and no quotation marks at all when we actually use them. This is because these variables are not in the formal language PROP (also called

Object Language); they *are* actually included among the enhancing symbols of our metalanguage M(L). Once it has been stipulated in the grammar that these metalinguistic variables stand for well-formed formulas, we no longer need to place them within quotation marks. It is like writing "Another name for Robert is 'Bob,'" but, subsequently, writing "Bob is a student" and not "'Bob' is a student."]

[Another device we will apply to avoid using special marks is this: we write the formal symbolic expression on a separate line by itself with an indent in front of it.]

- If "φ" and "ψ" are symbols of – not necessarily atomic – well-formed formulas (wffs), then:
  - $\sim \varphi$ and $\sim \psi$ are wffs.
  - $\varphi \cdot \psi$, $\varphi \vee \psi$, $\varphi \supset \psi$, and $\varphi \equiv \psi$ are all symbols of wffs.
  - Parentheses are used as auxiliary symbols *only* to prevent ambiguity. Liberalizing conventions are, therefore, permissible insofar as there is no risk of ambiguity.
  - Nothing else is a wff.

    [We absolutely need this mandate, which is known as the Closure Clause. Without it, our grammar is left wide open to arbitrary intrusions that will make it dissolve. A formal language is not an evolving organism – as the case is with the languages we speak and write in everyday life.]

We can agree, by means of a liberalizing convention, to omit external or outer parentheses insofar as there is no risk of ambiguity because of this. Moreover, conjunction and disjunction are commutative; this means that the following equivalences hold.

$$((\varphi_1 \vee \varphi_2) \vee \varphi_3) \equiv (\varphi_1 \vee (\varphi_2 \vee \varphi_3))$$

This means that shifting of parentheses does not change meaning – which is truth value for all specified assignments of truth values to the individual component variables. This renders the use of parentheses dispensable. This applies to the wedge and the dot. In contrast, the following two symbolic expressions are not logically equivalent – they do not always obtain the same truth values for assignments of truth values to their components. Thus, removal of parentheses generates ambiguity about meaning.

$((\varphi_1 \supset \varphi_2) \supset \varphi_3)$ is not equivalent with $(\varphi_1 \supset (\varphi_2 \supset \varphi_3))$

Thus,

$\varphi_1 \supset \varphi_2 \supset \varphi_3$

is ambiguous between $((\varphi_1 \supset \varphi_2) \supset \varphi_3)$ and $(\varphi_1 \supset (\varphi_2 \supset \varphi_3))$.

- Symbols not in the given formal language are not recognizable; they are considered as generating nonsense; well-formedness is forfeited.
- The number of right and left parentheses ought to be equal to each other. If this is not the case, there is ambiguity and the formula is not well-formed.
- We have allowed for the use of subscripts from the positive integers; this supplies a stock of individual variables that has the denumerable size of infinity (they can be placed in one-to-one correspondence with the natural numbers.)
- Metalinguistic symbols are not in the object language and their appearance in a formula renders that formula ill-formed.

0.1.1 EXERCISES

(1) Determine if the following symbolic expressions are well-formed (grammatically correct, well-formed-formulas or wffs) in accordance with the given PROP-grammar.

a. $\sim\sim\sim(p \equiv q)$
b. $p \,!\, q$
c. $\psi \cdot (q \vee \sim(q \cdot \sim p))$
d. $(p \supset \sim q) \vee \sim(p \supset \sim(q \vee \sim t)$
e. $p \veebar q$
f. $p \vee \sim q \supset r$
g. $\sim\sim(p_{11} \equiv p_{24})$
h. $\supset p \equiv q \sim r$
i. $(\sim p_1 \vee p_2) \cdot \sim(p_1 \supset (q_{33} \supset q_{34}))$
j. $\sim(p \supset (p \supset (p \supset (p \supset p))))$
k. $\sim p \rightarrow \sim q$
l. $\sim((\sim(p \equiv \sim q) \supset \sim(\sim q \cdot \sim p)) \vee \sim(p \vee \sim q)) \equiv \sim(p \equiv (p \equiv \sim p))$

(2) A method that we will not pursue in this text is the so-called axiomatic approach to the systematic formulation of a logic. Axiomatization has an intrinsic appeal due to the ancient progeny of the familiar Euclidean system of geometry; it is also a secret of the logician's trade that construction of axiomatic systems simplifies certain metalogical tasks – like proving that the logic can have a decision procedure that can derive all and only the right results. Even though we are not pursuing axiomatic systems, we present an example and exercise here, with a view to offering a fleeting flavor of how this approach to formal construction works. You should be able to tell that this is work done on the syntactic side.

Assume that we have two correctly formed strings of symbols from our symbolic resources in a language £, which are posited as no-proof or self-proving theses (they are called axioms). This is given by stipulation and it is to be understood that the strings are self-anointed as correct or that they derive-themselves (which is basically like saying that they derive themselves degenerately, without needing a derivation.) Every other grammatically correct string is a thesis of the language if and only if it is correctly derivable from lines all of which comprise axioms and/or already derived theses. Derivation rules are needed to this effect and such rules are specified. The grammar is also specified. Nothing else counts as a derivation or a thesis. The overall formal language so constituted is the theoretically conceived collection of all the derivable theses of the language. Some of the symbols may be interpreted by means of semantic connectives. If we do have semantic interpretations (systematic and specified unique mappings) of all the resources of Λ into a language ∫ so that all and only the theses of the first are the logical truths of the latter, we have a harmonization of the two languages. We can say that ∫ has an axiomatization; that Λ is sound and complete with respect to ∫.

Λ:: {⋈, ⋉, ⋊}//Well-Formed: if φ is well-formed then so is φ⋈, ⋈φ, φ⋉, ⋉φ, φ⋊, ⋊φ//
Axiom 1: (A1): ⋈⋉⋈
Axiom 2: (A2): ⋈⋉⋊⋊⋈
Derivation Rule1: (R1): ---⋈ ⇒ ---⋊⋈
Derivation Rule 2: (R2): ---⋊⋊⋈ ⇒ --- ⋉⋈
Example of a Derivation in Λ, with justification lines (theses, derivation rules and lines on which the rules have been applied.)

1. ⋈⋉⋈	A1
2. ⋈⋉⋊⋈	D1(1)
3. ⋈⋉⋊⋊⋈	D1(2)

Can you prove the following theses? (Failure to prove does not establish that no proof is available.)

a. ⋈⋉⋉⋈
b. ⋈⋉⋊⋊⋊⋉⋈
c. ⋈⋉⋉⋊⋈
d. ⋊⋊⋈⋉⋈
e. ⋈⋈⋈

Are the axioms of the system independent? This means: is it impossible to prove either one from the other by using the given derivation rules? If it is possible, they are not independent: the one that can be proven from the other is redundant. Is this a problem, do you think?

(3) What conventions could we introduce into our grammar so that we can economize on parentheses? As a clue, consider ranking the binding strength of the connective symbols.

(4) Logical investigations often revolve around subjects like validity, consistency and logical truth. Logics can be characterized by what is called the relation of logical consequence which can be defined as the pairs of sets of premises and conclusions following from those sets. This sounds remarkably like what we have called validity. Alternatively, logical systems can be characterized by the sets of their logical truths. It is a good question – and a rather advanced subject – whether those two modes of characterization coincide. The short answer at this point – considering that this is not an advanced exploration – is that this is not always the case, it is not guaranteed to be the case, but it does happen to be the case for the standard propositional logic. Try to show how validity and logical truth can be inter-defined. (As a clue, consider, for each argument form, generating an implicative formula with the conjuction of the premises as antecedent and the conclusion as consequent. Consult what has been covered under discussion of the Deduction Theorem.) We can also inter-define argument form validity with inconsistency. How would we do that? (Consider whether the conjunction of the premises and the negation of the conclusion are consistent or inconsistent in the case, and only in the case, of a valid argument form.)

(5) We have briefly examined only implication and equivalence as relations between propositions. Other relations can be defined. Two propositions are mutually independent if and only if neither one implies the other. How should we precise this concept precisely, so

that we can check, given a pair of propositional formulas, whether they are mutually independent? Other relations include the relation of mutual contrariety: two propositions are mutual contraries if and only if they cannot be both true together. Notice that we are not adding any other restriction besides "not being, possibly, true together." Choose which of the following can be the case if two propositions are mutual contraries.

a. They can both be true.
b. It can be that the first is true and the second is false.
c. It can be that the first is false and the second is true.
d. It can be that they are both false.

Finally, another relation is that of sub-contrariety. Two propositions are mutual sub-contraries if and only if they cannot be false together. From the above four options, which options apply in the case of sub-contrariety?

0.2 TRUTH TABLES

We call our formal notational idiom for the implementation of the truth table method PROPtt. The truth-tabular presentation of Propositional Logic – which can be used beyond the standard or two-valued case – is one of the most familiar and pleasantly simple subjects in the study of formal logic. The method is diagrammatic and replaces the matrix or some other similar construct. The truth table method is a decision procedure for Propositional Logic: it can be shown rigorously that the method both terminates in a finite number of steps and is guaranteed to provide correct decisions regarding such logical properties as validity of argument forms, logical status of propositional formulas and consistency of sets of propositional formulas. To state that we have a truth-tabular approach to a logic is to assert that this logical system has available to it a decision procedure. Depending on the number of atomic (individual, single) propositional variables in the formulas, the number of rows in the truth table can grow easily beyond our capacity to execute with our unassisted efforts but the metalogical proofs we have available show us that the truth-tabular method always turns out, in a finite number of steps, the correct decision. This is a logical calculator, answering to an ancient aspiration about implementation of some appropriate apparatus for the purpose of determining whether proffered arguments are valid or presented systems of opinions or theories are consistent. Moreover, determining that a proposition has the logical form of a tautology or logical truth, or of a contradiction or logical falsehood for that matter, is a mechanical procedure for the truth table but it is not immediately evident to the untrained rational agent; disputes about logical truths and logical falsehoods - essentially, pseudo-disputes - can be obviated in this way as much as invalid arguments and inconsistent theories can be rejected.

All of this is accessible but on the presupposition that our formalization or translation into the propositional symbolic idiom is adequate in the sense that it captures the logical structure of the propositions and argument forms that are translated. Unfortunately, there are limits to this. Moreover, the predicate logic, extending propositional, is ultimately bereft of a decision procedure. Infinitarian series rear their meddlesome heads in that case. Thus, the truth table does not carry over to - it is glaringly absent from - predicate logic systems. There is a possibility, however, of confining ourselves to finitarian models (which have domains with a finite number of objects), so that we might be able to use the truth table in such a limited case. Moreover, the connectives of predicate logic are the same in their definitions (hence, meanings) as those of Propositional Logic. Operations regarding strictly the connectives are computational and the truth table is a popular and handy method for defining the connectives and carrying out truth-functional computations for propositional formulas. Propositional Logic exemplifies the principle of Compositionality of (Logical) Meaning: if the truth values of the components of a propositional formula are assigned (specified), then the truth value of the entire propositional formula (corresponding to the truth value of its main-connective symbol) can be determined correctly. The truth table is like a map of all the possible valuations (assignments of truth values to the atomic component variables); hence, the truth table, in a mechanical and executable fashion, presents all the logically possible truth values which the given formula or set of formulas can obtain.

The truth table is a semantical device. It receives inputs into it and turns out outputs, all of which are truth values. The inputs are given or assigned to the individual or atomic letters (propositional variables) of the given formula or formulas. These are all the logically possible valuations or value-assignments (also called cases, interpretations and options.) The outputs are truth values for the given formula or formulas. Starting from the left, we construct columns for each one of the individual letters. Continuing to the right, we have the formulas themselves. In the case we evaluate an argument form, one of the given formulas is presented as distinguished (as the conclusion); this may be separated from the rest – as the last one to the right – by means of some special symbol. Such symbols, and other convenient symbolic and diagrammatic arrangements that are superimposed on the truth table – should be considered metalinguistic. Every row of the truth table can be understood by means of a convenient narrative – another indication that this approach is semantical. We can think of every row as a logically possible world. Such worlds are deprived of content-information – as one should expect since deductive reasoning is extra-empirical and depends on structures and, deeply, on the mere definitions of the logical connectives or operators. Each logically possible world is considered as completely described in the sense that every propositional variable – and, computationally, every compound formula – receives a determinate truth value in that possible world. Thus, these are worlds whose characterization is logical, not physical or factual. We can think of each logically possible world as labeled by the combinations of truth values given to the atomic propositional variables there. If we interpreted the atomic variables by means of linguistic propositions (for instance, "there is exactly one moon revolving around our planet"), then, taking into consideration all the other interpretations, one of the logically possible worlds exactly would correspond to our actual world. This is one row of a huge truth table - which we will not need to construct, but this is used for illustration purposes. Notice, however, that this row – the actual world row – is no more privileged than any other row. From a logical point of view, the consistent description of our world is just one of all the logically possible world-states descriptions. To study deductive logic characteristics, however, we are interested in what happened in all, or in some, or in none, of the logically possible worlds. For instance, a logical truth is true, by definition, in all logically possible worlds. An invalid argument has all premises true and its conclusion false in at least one logically possible world (notice that this might not be our actual world, again...)

Our grammar of PROPtt uses the same symbols and grammatical arrangements as the formal idiom PROP we constructed above. The truth table is built of rows and columns. It is crucial that we know the algorithm by which we determine the number of rows our truth table has, so we can proceed to constructing it. If the number of the kinds of atomic letters is n for all the given well-formed formulas, then the truth table has 2^n rows. Of course, the number of columns is the number of the individual letters (the input columns) plus the number of the given formulas (the output columns.) If other subdivisions are applied – to conveniently assist with computations – those are to be considered as metalinguistic superimpositions on the truth-tabular diagram. Here is an example of a truth table of one given well-formed formula (wff). The given formula has exactly 2 kinds of atomic letters (regardless of the fact that it has two occurrences of each letter – marking the occurrences from left to right if we are to label them.) Thus, the truth table has the number of rows:

$$2^n = 2^4 = 4$$

The promised algorithm for building the truth table – one of many possible such recipes – ensures that no input-combination is omitted. We start from the right-most input column and write, from the top to bottom, the truth values alternately as T and F. We then move to the left – the next input column – and write the truth-value inputs by two's as T-T-F-F from top to bottom. The next column to the left – if one were needed – would have the truth values written by four's as T-T-T-T-F-F; and so on. We can account for the fact that the number of input values from top to bottom for every column is a multiple of 2 since the number of rows is an exponentiation of 2.

The truth table can be deployed to give the definitions of the connectives of our formal language. Obviously, the truth table can give us all mathematically definable connectives, including those we do not include officially in our formal idiom. Here are the truth-tabular definitions along with the names of the symbols and the names of their denoted functions.

p	~p	pq	p·q	p∨q	p ⊃ q	p ≡ q
T	F	TT	T	T	T	T
F	T	TF	F	T	F	F
		FT	F	T	T	F
		FF	F	F	T	T

~: tilde: Negation.
·: dot: Conjunction.
V: wedge: Inclusive Disjunction.
⊃: horseshoe: (Material) Implication or (Material) Conditional.
≡: (Material) Equivalence or (Material) Biconditional.

Names of Joined Molecular Components by Connective:
~: negated propositional formula.
·: conjuncts.
V: inclusive disjuncts.
⊃: antecedent (to the left of the horseshoe), and consequent (to the right of the horseshoe.)
≡: (material) equivalents (mutual equivalents.)

These connectives interpret semantically Boolean functions over the set {1, 0}. The defining valuations can be presented metalinguistically as follows.

√(p) = T or F (Principle of Excluded Middle)
Not (√(p) = T and √(p) = F) (Principle of Non-Contradiction)

√(~ φ) = T if and only if √(φ) = F; √(~ φ) = F otherwise. [For any well-formed formula, wff, φ.]

√(φ · ψ) = T if and only if √(φ) = √(ψ) = T; √(φ · ψ) = F otherwise.

√(φ V ψ) = F if and only if either √(φ) = F and √(ψ) = F; √(φ V ψ) = T otherwise.

√(φ ⊃ ψ) = F if and only if √(φ) = T and √(ψ) = F; √(φ ⊃ ψ) = T otherwise.

√(φ ≡ ψ) = T if and only if √(φ) = √(ψ); √(φ ≡ ψ) = F otherwise.

Nothing else counts as a definition of a connective for PROP.

Thus, a logical connective of Propositional Logic is a complete function (called Truth Function), from domain to range as shown below for unary (one-place, monadic), binary (two-place, dyadic) and n-adic or n-place connectives.

$\int^{1}(p): \{T, F\} \longrightarrow \{T, F\}$
$\int^{2}(p): \{<T, T>, <T, F>, <F, T>, <F, F>\} =$
$\{T, F\} \times \{T, F\} \longrightarrow \{T, F\}^{2} \longrightarrow \{T, F\}$
$\int^{n}(p): \{T, F\}^{n} \longrightarrow \{T, F\}$

The range is always the set of truth values. [For Set Theory concepts and terms see chapter V-Appendix on Set Theory.] The domain is some Cartesian product of the set of truth values (starting with the degenerate case of the product of the set of truth values by itself and all the way to the nth product.)

An example of the truth table of a well-formed formula is shown below.
$\varphi = (p \equiv \sim q) \supset (p \cdot q)$

	p	q	(p	≡	~	q)	⊃	(p	·	q)
1	T	T	T	F	F	T	**T**	T	T	T
2	T	F	T	T	T	F	**F**	F	F	F
3	F	T	F	T	F	T	**F**	T	F	T
4	F	F	F	F	T	F	**T**	F	F	F
	1	2	3	4	5	6	7	8	9	10

The input columns are 1 and 2 (labeled by the metalinguistic numeral values at the bottom – which do not constitute a row of the truth table itself!) The output columns are 3-10. The main connective symbol – the largest-scope connective symbol – is on column 7. Accordingly, the truth value of the given formula is the value at column 7. The input rows provide all the logically possible (mathematically available) combinations of truth values for the atomic components. If we symbolize a value-assignment by "√", we have:

$$\surd = \{<T, T>, <T, F>, <F, T>, <F, F>\}$$

We can say that for the 2nd row, for instance, we have:

$$\surd(p) = T \text{ and } \surd(q) = F.$$

The given well-formed formula takes output values that are both Ts and Fs. It is logically indeterminate – to be called, by a more common nomenclature, logically contingent. It is not a logical truth (tautology) since it does not take T at every row (every logically possible case) and it is not a logical falsehood (contradiction) since it does not take the value F for every row.

The truth table as a presentation method and decisional procedure cashes on the computationality of the propositional system. Consider, for instance, the following example of computation of the truth values of given formulas for specified value-assignments to the ultimate individual components.

$\surd(p) = T, \surd(q) = F, \surd(r) = F.$

p	q	r	$p \supset (q \equiv \sim r)$	$\sim (p \vee (q \cdot r))$
T	**F**	**F**	**T** F F F T F	**F** T T F F F

This computation represents a row of the truth table for the two formulas. The truth table, of course, presents all the logically possible valuations.

Shortcut computations are possible based on considerations captured by the following metalinguistic equations. Looking at the truth tables that define the connectives, we can readily establish that all these equations can be justified. If one of the disjuncts is true, that suffices for the disjunction formula to be true. If one of the conjuncts is false, that suffices for the conjunction formula to be false. If either the antecedent is false or the consequent is false in an implicative formula, that suffices for the formula to be true. Moreover, all formulas expressing tautologies (logical truths) are true and all formulas expressing logical contradictions (logical falsehoods) are false but such discernment is not available to the casual or beginning student – and, indeed, it might not be available at all to mere inspection. Furthermore, given that certain relation obtain among logical truths and falsehoods, we can draw other conclusions about formulas: for instance, because every tautology is logically equivalent with every other tautology, a formula that connects two tautologies by the logical equivalence connective sign is always true. These and other similar considerations allow us sometimes to determine the truth value of a given formula even if we don't have all the values of the component atomic letters.

<u>Can be computed without knowing all the values</u>

$\surd(\varphi \vee T) = \surd(T \vee \varphi) = T$ $\quad$ $\surd(\varphi \cdot F) = \surd(F \cdot \varphi) = F$

$\surd(\varphi \supset T) = \surd(F \supset \varphi) = T$ $\quad$ $\surd(\varphi \supset \varphi) = \surd(\varphi \equiv \varphi) = T$

$\surd(\varphi \equiv \sim \varphi) = \surd(\sim \varphi \equiv \varphi) = F$

$\surd(\varphi \vee \sim \varphi) = \surd(\sim \varphi \vee \varphi) = T$

$\surd(\varphi \cdot \sim \varphi) = \surd(\sim \varphi \cdot \varphi) = F$

<u>The following cannot be computed without the values</u>

$\surd(\sim \varphi \supset \varphi) = \surd(\varphi)$ $\quad$ $\surd(\varphi \supset \sim \varphi) = \surd(\sim \varphi)$

The truth table system is mechanical. No ingenuity is required for implementing the method. The truth-tabular method, as a decision procedure, is correct – or, as it is called, sound – with respect to the semantics of Propositional Logic: every decision it returns is what should be. Conversely, in every case in which a property fails to apply, the truth-tabular method allows for that semantic fact to be recorded. In the case of checking for argument-form validity, this means that the truth table yields counterexample values – the truth value assignments for the atomic variable letters, for which all the premises are true and the conclusion is false.

The truth table can be used to determine validity and invalidity of argument forms. The full truth table for the formulas (premise formulas and conclusion formula) shows the rows, if any, for which all the premises are true and the conclusion is false. Such a row, if it is present, presents the counterexample values. If, and only if, no such row is present, then the given argument form is valid: it is not invalid (since no counterexample can be produced), and, of course, the bivalent character of our Propositional Logic ensures that invalidity and validity are the only possibilities. Examples are shown below - including about how to produce a partial truth table. This latter option consists in producing less than all the formula outputs insofar as it can still be determined whether the given argument form is valid or not. Reflecting on the fundamental definitions of logic, we can explain how this can be done. Detecting invalidity is the target - since validity is automatically determined just in case invalidity does not obtain. Only if there is a counterexample do we have an invalid argument form. Any counterexample row, by definition, must have a false conclusion. Thus, we may start with computing only the conclusion column outputs and limit our further inquiries only to the rows that have a false conclusion. Additional simplifications are possible, resulting in further reductions of the truth table: across the rows with false conclusion we can immediately disregard any row for which any false premise appears. This row cannot possibly yield a counterexample since it cannot have *all* premises true - even though it has a false conclusion. So, we can move on and economize considerably on labor.

	p	q	p ⊃ q	p ⊃ ~ q	/.. ~p
1	T	T	**T**	**T F F**	**F**
2	T	F	**F**	**T T T**	**F**
3	F	T	**T**	**F T F**	**T**

4	F	F	T	F T T	T
			1	2 3 4	5

The column labeled 1 gives us the values of the first premise for the characteristic valuations of the input rows. The truth value of the second premise is on column 3. Column 5 is the conclusion output column. The formula outputs for each row are:

Row 1: <T, F, F> -- valuations: $<p^T, q^T>$
Row 2: <F, T, F> -- valuations: $<p^T, q^F>$
Row 3: <T, T, T> -- valuations: $<p^F, q^T>$
Row 4: <T, T, T> -- valuations: $<p^F, q^F>$

This is a valid argument form. There is no row in which all the premises are true and the conclusion is false – no counterexample row. We see below the partial truth table we can deploy for assessment of the given argument form. We only compute output values for the rows for which the conclusion formula obtains a false value.

	p	q	p ⊃ q	p ⊃ ~ q	/.. ~ p
1	T	T	**T**	**T F T**	**F**
2	T	F	**F**	**T T T**	**F**
3	F	T			**Ŧ**
4	F	F			**Ŧ**
			1	2 3 4	5

The only relevant rows are those labeled as 1 and 2.

There is an intuitive ring to claiming validity for the expression of this argument form in ordinary language: if a proposition implies both some other proposition and the negation of that proposition, then our given proposition must be false. Implication and validity are harmonized in the standard Propositional Logic: φ implies ψ if and only if the implicative proposition φ ⊃ ψ is a tautology. Thus, we can restate the above: if assuming a proposition we can prove both some proposition and that proposition's negation, then our assumed proposition must be false (cannot be true.) If we can prove both a proposition and its negation, then we can prove their conjunction – which is a contradiction. Thus, we can say: if we can prove a contradiction from a given proposition, then our given proposition must be false.

The truth table method can also be applied to check for consistency of a given set of well-formed formulas. We produce the truth table and check if there is any – at least one, even one – row across which all the formula values are true. This is what we mean by consistency: the logical possibility that all the propositional formulas are true together. Although a term like "possible" is rather recalcitrant in the sense that it resists ordinary intuitions, the truth table shows us all the logical possibilities. A row represents meanings (truth values) that characterize a logical possibility. Thus, a row with all true outputs shows that there is, indeed, a logically possible case in which all the given formulas are true together.

Finally, the truth table can be deployed to assess the logical status of a given single formula and also what relationship obtains between a given pair of propositional formulas. The three types of logical status for a propositional formula are: tautology (logical truth, necessary truth, validity), contradiction (logical falsehood, necessary falsehood, invalidity), and contingency (indeterminate formula, indefinite formula, possible truth and possible falsehood.) We can express – and, indeed, correctly provide alternative definitions for – these concepts by appealing to the truth table. We can do this as follows.

A tautology is a formula whose truth table has all output rows as true. A contradiction is a formula whose truth table has all output rows as false. A contingency is a formula whose truth table has both true and false values (regardless of how many of each) for its output values.

	p	q	(p	⊃	~	q)	⊃	(q	⊃	~p)
1	T	T	T	F	F	T	**T**	T	F	F T
2	T	F	T	T	T	F	**T**	F	T	F T
3	F	T	F	T	F	T	**T**	T	T	T F
4	F	F	F	T	T	F	**T**	F	T	T F
	1	2	3	4	5	6	7	8	9	10

This formula expresses, within the PROPtt language, a tautology of Propositional Logic.

The case of relations between two propositions includes logical implication, logical equivalence and, of course, consistency which we have already defined for the case of two or more propositional formulas. Two formulas are logically equivalent if and only if their truth table presents exactly the same truth values for the two formulas on every row. Or, if we form the formula that is generated by placing the triple bar (equivalence symbol) between the given formulas – the formula so produced checks as a tautology. Next, we define implication. Given two formulas, the first implies the second if and only if there is no row of their truth table for which the first receives true and the second receives false; the second implies the first if and only if there is no row of their truth table for which the first receives false and the second receives true as truth value.

Informally, we can expedite our computational adventures significantly by using a method that is called Short Computational Method. We can also call it Quick Truth Table as we notice that it consists in something like constructing a maximally shortened partial truth table – one with one or more, but very few, rows. This approach is not officially included in the PROPtt idiom, although we poach its symbols and grammar to enact executions of the method within our metalanguage. We can use this method to check for validity – and we can leave it as an exercise how to use this method also for the other purposes to which we apply the truth table.

We pretend to be laying out a counterexample row of the given argument form's truth table. Of course, we don't know that this argument form is invalid - and, hence, we don't know that it has a counterexample. We should be able to distinguish success from failure upon termination of the procedure - and we assume, as we can also show, that the method always reaches a point of termination. Obviously success means that we have indeed generated a counterexample-row; thus, the form is invalid. Failure, on the other hand, means that no counterexample-row is available; this is because our pretense is for the generic case - we make no special accommodations or impose restrictions on our endeavor and we cover all possibilities for which we have all true premises and a false conclusion; thus, our failure is a failure of any attempt to produce a counterexample. We will see how failure is determined.

Since we are trying to generate a counterexample row, we assign true to each premise and false to conclusion. From that point on, we figure out what other assignments we can make. You may recall the equations given earlier, aiming how to assist with computations in cases in which we do not have all the atomic valuations. If we are unfortunate, we might have to add more than one rows as we cover all possible cases. Still, this is the most we can do with a view to using a partial instead of a complete truth table. As we indicated, it is important to define "failure" in this process. We stipulate that we continue assigning values across the row (or rows) until saturation (until all formulas have received values.) Failure, then, consists in discovering that at least one symbol receives both truth values, true and false. We show an example below, proceeding in phases as we explicate the process. We label each phase and, once we reach valuations for the atomic variables, we indicate that next to the label.

	p	⊃	~	q	/∴	q	⊃	~p
1.		T					F	
2. <q^T>						T	F	F
3. <p^T, q^T>		T		T		T	F	F T
4. <p^T, q^T>	T	T		T		T	F	F T
5. <p^T, q^T>	T	**T**	F	T		T	F	F T
6. <p^T, q^T>		**F**						

1. We assign true to the premise and false to the conclusion.
2. The conclusion is an implicative formula. The only way for it to be false is if the antecedent is true and the consequent is false. This gives us the truth value of⌜ q ⌝in 3.
4. We obtain the value of ⌜ p ⌝ since we have its negation as false; it must, then, be true. Thus, we have both truth values for the atomic variables for this suppositious row we are constructing.
5. We compute the value of the consequent for the premise.
6. The premise is an implicative formula; thus, it obtains the value false on this row. Recall, however, that we are presuming it to be true for the sake of obtaining a counterexample. Thus, we are compelled to assign both true and false to some symbol in the formulas and this constitutes failure of the effort. Since it is impossible to produce a counterexample, our given argument form is valid.

We do not have a symbolic language for predicate logic yet. Overall, the truth table cannot be used in predicate formalisms. It is still a useful device for grasping fundamental logical notions – like validity, consistency and the rest.

0.2.1 EXERCISES

(1) The following are all tautologies of Propositional Logic. Use the truth table decision procedure to verify that.

a. $\vdash \sim (p \supset q) \supset \sim q$
b. $\vdash \sim (p \vee q) \equiv (\sim p \cdot \sim q)$
c. $\vdash \sim (p \cdot q) \equiv (\sim p \vee \sim q)$
d. $p \supset q \vdash \sim (\sim p \supset q)$

e. $\vdash ((p \supset q) \supset p) \supset p$
f. $\vdash (p \supset (p \supset q)) \supset (p \supset q)$
g. $\vdash ((p \supset q) \supset q) \supset ((q \supset p) \supset p)$
h. $\vdash ((p \supset q) \supset r) \equiv ((\sim p \supset r) \cdot (q \supset r))$
i. $\vdash (p \supset (q \supset r)) \equiv ((p \cdot q) \supset r)$
j. $\vdash (p \vee (q \cdot r)) \equiv ((p \vee q) \cdot (p \vee r))$
k. $\vdash (p \cdot (q \vee r)) \equiv ((p \cdot q) \vee (p \cdot r))$

Use the truth table to determine logical status, consistency of formulas, and validity of argument forms. You can tell what is expected because only argument forms have characteristic separation of the conclusion formula from the premises formulas; many formulas given together are to be checked for consistency and one solitary formula is to be checked for its logical status (tautology, contradiction or contingency.) Of course, completing the full truth table in any case renders the status of all the formulas perspicuous in all cases.

a. $\{(p \vee \sim (q \equiv \sim (p \supset \sim (p \cdot \sim q)))), \sim (p \cdot \sim (p \vee \sim q))\}$
b. $p \supset (p \supset q), \sim (p \equiv \sim (p \supset q)) \vdash q \vee p$
c. $\sim p \supset (\sim q \cdot r), p \equiv (q \cdot r) \vdash \sim p \supset \sim (q \cdot \sim r)$
d. $\sim \sim (p \vee q \vee (\sim q \vee p)))$
e. $((p \equiv (q \vee \sim p)) \cdot \sim (p \supset (q \cdot \sim p))) \vdash \sim (p \vee \sim q)$
f. $\sim (\sim s \equiv (s \supset (s \supset \sim t))) \cdot (p \supset (\sim p \supset (s \vee t)))$
g. $\{\sim p \supset (\sim s \equiv \sim t\}, t \vee \sim (p \cdot \sim s), \sim p \supset \sim t\}$
h. $\sim p \vee \sim (q \cdot r) \vdash (p \supset (q \supset r)) \equiv ((p \cdot q) \supset r)$
i. $p \supset t, q \supset t, \sim ((p \cdot q) \supset t) \vdash p \vee q \vee t$
j. $\sim (p \cdot (q \supset (\sim q \equiv \sim r)))$

(2) Determine by use of the truth table procedure if the following sets of formulas are consistent or inconsistent.

a. $\{\sim p \vee \sim (q \cdot r), q \supset (\sim p \equiv q), \sim q\}$
b. $\{\sim (p \cdot (q \supset \sim r)), \sim p, \sim r, ((p \vee q) \equiv (\sim r \supset \sim q))\}$
c. $\{p \supset (\sim p \supset (p \supset \sim p)), ((q \supset \sim q) \supset q) \supset \sim q, (p \vee \sim q) \cdot (\sim p \vee q)\}$
d. $\{\sim (p \equiv \sim q), p \cdot q, \sim p \vee q, \sim p \supset (\sim q \supset \sim p)\}$
e. $\{\sim p \vee (\sim q \supset (\sim r \supset \sim p)), \sim \sim (p \supset \sim (q \cdot r))\}$

f. $\{(p \supset \sim q) \supset (\sim p \equiv r), (\sim p \supset q) \supset (p \equiv \sim r)\}$
g. $\{\sim p \cdot \sim q \cdot \sim r, (p \vee q) \supset (q \vee r), (p \vee r) \supset (q \vee r)\}$
h. $\{\sim (p \supset (p \supset \sim q)), \sim p \supset \sim (q \supset \sim q), q \supset (\sim p \supset \sim q)\}$

(3) Use the Short Computational Method to determine if the following argument forms are valid or not.

a. $(p \supset q) \supset ((q \supset p) \supset p)$
b. $(p \cdot q) \supset (p \equiv (q \vee \sim p))$
c. $((p \supset q) \equiv q) \supset \sim p$
d. $p \supset (\sim q \supset (q \supset \sim p))$
e. $p \equiv (q \equiv (r \vee (\sim r \cdot p))$
f. $\sim (p \supset (q \supset \sim (r \vee p))$
g. $(p \vee (q \supset \sim r)) \supset ((p \cdot (q \vee r))$
h. $\sim \sim p \vee \sim (q \supset \sim (q \cdot \sim r))$

(4) Can we use the Short Computational Method to check for consistency or for the logical status of a given formula? If so, how?

(5) Refer to the truth table to justify answers to the following questions.

a. Can we tell if an argument form with inconsistent premises is valid or invalid?
b. Can a tautology be consistent with a contradiction?
c. Can two logically contingent formulas be consistent? Do they have to be?
d. Can we tell if an argument form with a tautology as conclusion is valid or invalid?
e. Can we tell if a set of logically contingent premises and a logically contingent conclusion is valid or invalid?
f. What is the negation of a logical contingency?
g. Does the inclusive disjunction of two logical contingencies have to be a logical contingency?
h. Does the negation of the conjunction of two tautologies have to be a contradiction?

i. Why is one counterexample row sufficient to check the argument form as invalid?
j. Why is it sufficient to check only the rows of a truth table in which the conclusion is false?
k. Formulate argument validity and invalidity in terms of consistency or inconsistency between the conjunction of the premises and the negation of the conclusion.

(6) Having practiced the full truth table regimen, now return to the exercises above and apply the partial truth table mechanism.

0.3 A SYSTEM FOR NATURAL DEDUCTION: PROPND*

Proof-theoretic systems for Propositional Logic are syntactical in their inception and execution: we engage in manipulation of symbols and execution of actions, as regulated by rules, rather than in building models and assigning truth values under specified conditions – as the case is with the semantic approach. We will construct a system, PROPnd*. Since our purpose in this text is not specifically the study of Propositional Logic, we opt for a conveniently user-friendly system like those one finds in elementary textbooks. We specify the grammar and we avail ourselves of metalinguistic symbolic resources as indicated. Our system will have redundancy in its system of rules: some of those derivation rules can be dispensed with and yet we can prove all the theses that are provable by using the other rules in the system. The objective is to make proofs easier.

As the case is with natural deduction systems in general, failure to find a proof does not mean that no proof is in-principle available in the system – it could well be a matter of lapses in ingenuity. Thus, this system is not a mechanical computational or diagrammatic proof method whose termination always ensures success or, in the case of failure, decides values for a counterexample. On the other hand, success in our natural deduction system means that the proven formula expresses a thesis or theorem of the standard Propositional Logic. Although the system PROPnd* has redundancy of rules in the sense indicated above, it does have all the rules needed (and then some) for proving all provable theses of the standard Propositional Logic.

As a pragmatic virtue – conducing to its easy application – our system exacts a price: it hides from the theoretical logician certain important features. We do not have exactly the number of rules that suffices for proving all, and only, the provable theses of Propositional Logic; as we said, we have more than that. This may create an impression that the standard Propositional Logic is profligate in its proof-theoretic requirements, which is not the case. None of our derivation or deduction rules are presented as introduction or elimination rules for connectives: although this might seem trivial at first, there are significant theoretical implications – and broader philosophical stakes – in the deployment of introduction and elimination rules for connectives. We ignore such issues in the present text.

GRAMMAR (PROPnd*):
{A, B, ...} (for translation from English, with subscripts from the natural numbers set, N)
{p, q, ...} (atomic sentence variables in logical forms - with subscripts from N)
{φ, ψ, ...} (symbols for well-formed formulas, not necessarily atomic - with subscripts from N)
{∏, ∑, ...} (symbols for well-formed formulas in rule schemata - subscripts from N)
{~, ·, ∨, ⊃, ≡} (connective symbols)
{(,)} (auxiliary symbols - parentheses to be used only for disambiguation)

We write the inferential or derivation rule schemata by using different symbols from those used in the object language of PROPnd*. The rules are to be thought of as recipes giving instructions for systematic action. The rules have metalinguistic symbols for variables (not necessarily simple) from the set {∏, ∑, ...}, and possibly subscripts from the set of positive integers. Other symbols – for connectives and the auxiliary parentheses symbols - are the same as in the object language - something that should be considered as a matter of accidental sharing of symbols. The student of the natural deduction system should be able to discern instances of applications of the rules in our object language; similarly, the student of the system should be able to apply rules to move from one line to the next. Practice is needed for this purpose; we relegate this task to the exercises.

In the metalanguage in which the derivation-rules schemata are written, we use "⊢" to indicate one-directional derivation license (from left to right) and "⊣⊢" to indicate two-directional derivation license (either from left to right or from right to left.) A formal proof - as encapsulated in the generic derivation schema shown below - is the complete collection of lines, positioned vertically from top to bottom; the premises are designated; assumed premises (also called posited premises) are also designated as such. Every line besides those for premises and assumed premises requires justification - which is provided to the right and is considered as metalinguistic in its mode of transaction. Justification is given by means of reference to the label or name of the deduction rule that has been applied and the number(s) of the line(s) on which the rule has been applied.
Premises are self-justifying. Assumed premises do not need justification but they impose a burden or incur a cost. The proof cannot be considered terminated unless all the assumed premises have been discharged. The discharge should proceed from the last assumed premise (which should be the first to be discharged) to the first assumed premise.
Because we are using a simpler system, we only run into the use of assumed premises - and the attendant burden of discharge - with respect to two proof-methods, which we include in PROPnd*, and which we will call Conditional Proof and Indirect Proof. Accordingly, the columns for used and discharged premises can be dispensed with unless Conditional Proof (CP) or Indirect Proof (IP) is used. It is permitted that both CP and IP can be deployed for a proof.
The last line, when the proof is properly terminated, is the conclusion

as given in the sequent that is to be proven.

Sequent Proven: $\prod_1, \prod_2, \ldots, \prod_n \vdash \Sigma$

Proof Lines	wffs	Justification	Premises Used	Premises Discharged
1.	$\prod_1$	P1 or $AP1^-$	P1 / $AP1^-$	
2.	$\prod_2$	P2 or $AP2^-$	P2 / $AP2^-$	
3.	---	--- ---	--- ---	---
4.	$\prod_n$	Rule R(line #, ...)	Pj / APi^-	---
5.	---	--- ---	--- ---	---
n.	Σ	Rule R*(line ^, ...)	--- ---	$AP1^0$, ...

The above rules are structural as they regulate the manner in which the proofs are conducted and properly terminated. Subsequently, we have connectives rules – and, as indicated above, a superabundance of them. Some of the rules are one-directional and others are two-directional (also called Replacement Rules.) One-directional rules can be applied to move downward on the proof-column. Their labels are written as justifications and their application is on lines that have preceded on the proof column. On the other hand, two-directional or replacement rules have the powerful characteristic that they permit substitution of parts within a formula. This reflects the extensional character of classical Propositional Logic as we can easily see from repairing to the definition of extensionality and the so-called principle of extensionality. If we cast a glance on the semantic side, we can define the Extensionality Principle as a license for mutual substitution of logically equivalent components of propositions without loss of truth-value for the compound proposition within which the substitution takes place.

■ PROPnd* Derivation Rules

If we think of the semantic translations of the formulas of PROPnd*, we should expect and request that all the deductive rules we have in our system correspond – in their semantic interpretations – to valid argument forms of the Propositional Logic. For two-directional rules, the interpreting arguments forms should be valid in both directions. The truth-table method can be put to use to verify this. We can think of the semantic translation as justifying the corresponding deduction rules. The names of the rules of PROPnd* are the same as the abbreviatory names for the justifying semantic sequents shown below. We symbolize by "⊢" the relation of logical consequence of the semantic interpretation of the standard Propositional Logic.

□ Semantic Correspondents of
One-Directional Inferential Rules (and abbreviations of rules' names)

1. Simplification (S) $\prod \cdot \sum \vdash \prod$ $\prod \cdot \sum \vdash \sum$
2. Conjunction (Conj) $\prod, \sum \vdash \prod \cdot \sum$ $\sum, \prod \vdash \prod \cdot \sum$
3. Modus Ponens (MP) $\prod \supset \sum, \prod \vdash \sum$
4. Modus Tollens (MT) $\prod \supset \sum, \sim \sum \vdash \sim \prod$
5. Addition (Add) $\prod \vdash \prod \vee \sum$ $\prod \vdash \sum \vee \prod$
6. Disjunctive Syllogism (DS) $\prod \vee \sum, \sim \prod \vdash \sim \sum$ $\prod \vee \sum, \sim \sum \vdash \sim \prod$
7. Constructive Dilemma (CD) $\prod 1 \vee \prod 2, \prod 1 \supset \sum 1, \prod 2 \supset \sum 2 \vdash \sum 1 \vee \sum 2$
8. Hypothetical Syllogism (HS) $\prod 1 \supset \prod 2, \prod 2 \supset \prod 3 \vdash \prod 1 \supset \prod 3$

□ Two-Directional (Replacement, Substitution) Inferential Rules (and abbreviations of names)

Double Negation (DN) $\prod \dashv\vdash \sim \sim \prod$
Contraposition (Contra) $\prod \supset \sum \dashv\vdash \sim \sum \supset \sim \prod$
Commutation of ⌜ · ⌝ (Comm·) $\prod \cdot \sum \dashv\vdash \sum \cdot \prod$
Commutation of ⌜ ∨ ⌝ (Comm∨) $\prod \vee \sum \dashv\vdash \sum \vee \prod$
Association of ⌜ · ⌝ (Assoc·) $\prod 1 \cdot (\prod 2 \cdot \prod 3) \dashv\vdash (\prod 1 \cdot \prod 2) \cdot \prod 3$
Association of ⌜ ∨ ⌝ (Assoc∨) $\prod 1 \vee (\prod 2 \vee \prod 3) \dashv\vdash (\prod 1 \vee \prod 2) \vee \prod 3$
Idempotence of ⌜ · ⌝ (Idemp·) $\prod \cdot \prod \dashv\vdash \prod$
Idempotence of ⌜ ∨ ⌝ (Idemp∨) $\prod \vee \prod \dashv\vdash \prod$
Distribution of ⌜ · ⌝ over ⌜ ∨ ⌝ (Dist·/∨) $\prod 1 \cdot (\prod 2 \vee \prod 3) \dashv\vdash (\prod 1 \cdot$

$\prod 2) \vee (\prod 1 \cdot \prod 3)$
Distribution of ⌜ ∨ ⌝ over ⌜ · ⌝ (Dist∨/·) $\prod 1 \vee (\prod 2 \cdot \prod 3) \dashv\vdash (\prod 1 \vee \prod 2) \cdot (\prod 1 \vee \prod 3)$

DeMorgan's Laws (DeM) $\sim(\prod \cdot \Sigma) \dashv\vdash \sim\prod \vee \sim\Sigma$
$\sim(\prod \vee \Sigma) \dashv\vdash \sim\prod \cdot \sim\Sigma$
Implication or Conditional Exchange (Impl or CE) $\prod \supset \Sigma \dashv\vdash \sim\prod \vee \Sigma$
Importation (Imp) $\prod 1 \supset (\prod 2 \supset \prod 3) \dashv\vdash (\prod 1 \cdot \prod 2) \supset \prod 3$
Exportation (Exp) $(\prod 1 \cdot \prod 2) \supset \prod 3 \dashv\vdash \prod 1 \supset (\prod 2 \supset \prod 3)$
Equivalence or Biconditional (Equiv) $\prod \equiv \Sigma \dashv\vdash (\prod \supset \Sigma) \cdot (\Sigma \supset \prod)$

Now we can show the deduction rules corresponding to the above valid semantic sequents in vertical pattern.

One-Directional Rules

·-Rules

Simplification (S)

k. $\prod \cdot \Sigma$
⋮
m. $\prod$ S(k)
⋮
n. Σ S(k)

Conjunction (Conj)

k. $\prod$
⋮
m. Σ
⋮
n. $\prod \cdot \Sigma$ Conj(k, m)

k. Σ
⋮
m. $\prod$
⋮
n. $\prod \cdot \Sigma$ Conj(k, m)

∨-Rules

Addition (Add)

k. $\prod$
⋮
m. $\prod \vee \sum$ Add(k)

Constructive Dilemma (CD)

k. $\prod_1 \vee \prod_2$
⋮
l. $\prod_1 \supset \sum$
⋮
m. $\prod_2 \supset \sum$
⋮
n. $\sum$ CD(k, l, m)

Alternative Schema – to facilitate shorter proofs.

k. $\prod_1 \vee \prod_2$
⋮
l. ↱$^1\prod_1$ AP1 (Assumed Premise 1)
m. ↱$^2\prod_2$ AP2 (Assumed Premise 2)
⋮
n. $\sum$ (... l...)
⋮
o. ↳^{1}↳$^2\sum$ (... m...)
p. $\sum$ CD(k, l, m, n, ..., o)

The two disjuncts are posited as assumed premises. But they represent a debt that is incurred; they have to be discharged, in the appropriate and specified way, for the proof to be considered terminated. The CD rule is used to discharge the two assumptions in the fashion indicated schematically above. If the conclusion is derivable from both discjuncts, then CD is applied to discharge the two assumed premises and to terminate the conclusion.

When we study Conditional Proof (CP) and Indirect Proof (IP) subsequently, we will come across other cases in which assumed premises are posited and need to be discharged by the application of the appropriate rule.

⊃-Rules

Modus Ponens (MP)

k. $\prod \supset \sum$
⋮
m. $\prod$
⋮
n. $\sum$ MP(k, m)

Modus Tollens (MT)

k. $\prod \supset \sum$
⋮
m. $\sim \sum$
⋮
n. $\sim \prod$ MT(k, m)

Hypothetical Syllogism (HS)

k. $\prod_1 \supset \prod_2$
⋮
m. $\prod_2 \supset \prod_3$
⋮
n. $\prod_1 \supset \prod_3$ HS(k, m)

Disjunctive Syllogism (DS)

k. $\prod \vee \sum$
⋮
m. $\sim \prod$
⋮
n. $\sum$ DS(k, m)

k. $\prod \vee \sum$
⋮
m. $\sim \sum$
⋮

n. ∏ DS(k, m)

#The DS-rule is often presented with a Latin name similar to that of Modus Tollens. So, we include it along with the horseshoe rules. When we turn to the Replacement Rules, we dispense with the categorization of rules in relation to connective symbols. A natural-deduction proof system for the standard Propositional Logic can be sufficiently constructed on the basis of MP, Add, CD, the DN-rule which is a replacement rule, and a rule (which we do not use) by which any variable symbol can be introduced once we have a <u>license</u> conferred by absurdity (which happens when any formula and the negation of the same formula appear on the proof lines.) We do not need an absurdity symbol for our present formal purposes. When we present the Indirect Proof method below, we may recall the above reference to the license-rule. All other rules are derivable. The proofs we can construct by using only the minimal set of rules mentioned above are often more cumbersome than the ones made possible within PROPnd*.

Two-Directional (Replacement) Rules

<u>Double Negation (DN)</u>

k. ~ ~ ∏
⋮
m. ∏ DN(k)

k. ∏
⋮
m. ~ ~ ∏ DN(k)

Replacement

k. ... ∏ ...
⋮
m. ... ~ ~ ∏ ... DN(k)

k. ... ~ ~ ∏ ...
⋮

m. ... $\prod$ ... DN(k)

<u>Contraposition (or Transposition) (Contra)</u>

k. $\prod \supset \sum$
⋮
m. $\sim \sum \supset \sim \prod$ Contra(k)

k. $\sim \sum \supset \sim \prod$
⋮
m. $\prod \supset \sum$ Contra(k)

Replacement

k. ... $\prod \supset \sum$ ...
⋮
m. ... $\sim \sum \supset \sim \prod$ ... Contra(k)

k. ...$\sim \sum \supset \sim \prod$ ...
⋮
m. ... $\prod \supset \sum$... Contra(k)

<u>Idempotence-· (Id·)</u>

k. $\prod \cdot \prod$
⋮
m. $\prod$ Idemp(·)(k)

k. $\prod$
⋮
m. $\prod \cdot \prod$ Idemp(·)(k)

Replacement

k. ... $\prod \cdot \prod$ ...
⋮
m. ... $\prod$ ... Idemp(·)(k)

k. ... $\prod$ ...
⋮

m. ... $\prod \cdot \prod$...   Idemp(·)(k)

<u>Idempotence(∨) (Id∨)</u>

k. $\prod \vee \prod$
⋮
m. $\prod$ Idemp(∨)(k)

k. $\prod$
⋮
m. $\prod \vee \prod$ Idemp(∨)(k)

Replacement

k. ... $\prod \vee \prod$ ...
⋮
m. ... $\prod$ ...   Idemp(∨)(k)

k. ... $\prod$ ...
⋮
m. ... $\prod \vee \prod$...   Idemp(∨)(k)

<u>Commutation(·) (Comm·)</u>

k. $\prod \cdot \sum$
⋮
m. $\sum \cdot \prod$ Comm(·)(k)

k. $\sum \cdot \prod$
⋮
m. $\prod \cdot \sum$ Comm(·)(k)

Replacement

k. ... $\prod \cdot \sum$ ...
⋮
m. ... $\sum \cdot \prod$ ...   Comm(·)(k)

k. ... $\sum \cdot \prod$ ...
⋮
m. ... $\prod \cdot \sum$... Comm(·)(k)

Commutation(∨) (Comm∨)

k. $\prod \vee \sum$
⋮
m. $\sum \vee \prod$ Comm(∨)(k)

k. $\sum \vee \prod$
⋮
m. $\prod \vee \sum$ Comm(∨)(k)

Replacement

k. ... $\prod \vee \sum$ ...
⋮
m. ... $\sum \vee \prod$ ... Comm(∨)(k)

k. ... $\sum \vee \prod$ ...
⋮
m. ... $\prod \vee \sum$... Comm(∨)(k)

Association(·)(Assoc·)

k. $\prod_1 \cdot (\prod_2 \cdot \prod_3)$
⋮
m. $(\prod_1 \cdot \prod_2) \cdot \prod_3$ Assoc(·)(k)

k. $(\prod_1 \cdot \prod_2) \cdot \prod_3$
⋮
m. $\prod_1 \cdot (\prod_2 \cdot \prod_3)$ Assoc(·)(k)

Replacement

k. ... $\prod_1 \cdot (\prod_2 \cdot \prod_3)$ ...
⋮
m. ... $(\prod_1 \cdot \prod_2) \cdot \prod_3$ ...Assoc(·)(k)

k. ... $(\prod_1 \cdot \prod_2) \cdot \prod_3$ ...
⋮
m. ... $\prod_1 \cdot (\prod_2 \cdot \prod_3)$ ...Assoc(·)(k)

Association(∨) (Assoc∨)

k. $\prod_1 \vee (\prod_2 \vee \prod_3)$
⋮
m. $(\prod_1 \vee \prod_2) \vee \prod_3$ Assoc(∨)(k)

k. $(\prod_1 \vee \prod_2) \vee \prod_3$
⋮
m. $\prod_1 \vee (\prod_2 \vee \prod_3)$ Assoc(∨)(k)

Replacement

k. ... $\prod_1 \vee (\prod_2 \vee \prod_3)$ ...
⋮
m. ... $(\prod_1 \vee \prod_2) \vee \prod_3$ ... Assoc(∨)(k)

k. ... $(\prod_1 \vee \prod_2) \vee \prod_3$ ...
⋮
m. ... $\prod_1 \vee (\prod_2 \vee \prod_3)$ ... Assoc(∨)(k)

Distribution(∨/·)(Distr∨/·)

k. $\prod_1 \vee (\prod_2 \cdot \prod_3)$
⋮
m. $(\prod_1 \vee \prod_2) \cdot (\prod_1 \vee \prod_3)$ Distr(∨/·)(k)

k. $(\prod_1 \vee \prod_2) \cdot (\prod_1 \vee \prod_3)$
⋮
m. $\prod_1 \vee (\prod_2 \cdot \prod_3)$ Distr(∨/·)(k)

Replacement

k. ... $\prod_1 \vee (\prod_2 \cdot \prod_3)$...
⋮
m. ... $(\prod_1 \vee \prod_2) \cdot (\prod_1 \vee \prod_3)$...

Distr(∨/·)(k)

k. ... $(\prod_1 \vee \prod_2) \cdot (\prod_1 \vee \prod_3)$...
⋮
m. ... $\prod_1 \vee (\prod_2 \cdot \prod_3)$ ... Distr(∨/·)(k)

<u>Distribution(·/∨)(Distr·/∨)</u>

k. $\prod_1 \cdot (\prod_2 \vee \prod_3)$
⋮
m. $(\prod_1 \cdot \prod_2) \vee (\prod_1 \cdot \prod_3)$ Distr(·/∨)(k)

k. $(\prod_1 \cdot \prod_2) \vee (\prod_1 \cdot \prod_3)$
⋮
m. $\prod_1 \cdot (\prod_2 \vee \prod_3)$ Distr(·/∨)(k)

Replacement

k. ... $\prod_1 \cdot (\prod_2 \vee \prod_3)$...
⋮
m. ... $(\prod_1 \cdot \prod_2) \vee (\prod_1 \cdot \prod_3)$...
Distr(·/∨)(k)

k. ... $(\prod_1 \cdot \prod_2) \vee (\prod_1 \cdot \prod_3)$...
⋮
m. ... $\prod_1 \cdot (\prod_2 \vee \prod_3)$ ... Distr(·/∨)(k)

<u>DeMorgan(DeM)</u>

k. $\sim (\prod \cdot \sum)$
⋮
m. $\sim \prod \vee \sim \sum$ DeM(k)

k. $\sim \prod \vee \sim \sum$
⋮
m. $\sim (\prod \cdot \sum)$ DeM(k)

k. $\sim (\prod \vee \sum)$
⋮
m. $\sim \prod \cdot \sim \sum$ DeM(k)

k. $\sim \prod \cdot \sim \sum$

⋮
m. $\sim(\prod \vee \sum)$ DeM(k)

Replacement

k. … $\sim(\prod \cdot \sum)$ …
⋮
m. … $\sim\prod \vee \sim\sum$ … DeM(k)

k. … $\sim\prod \vee \sim\sum$ …
⋮
m. … $\sim(\prod \cdot \sum)$ … DeM(k)

k. … $\sim(\prod \vee \sum)$ …
⋮
m. … $\sim\prod \cdot \sim\sum$ … DeM(k)

k. … $\sim\prod \cdot \sim\sum$ …
⋮
m. … $\sim(\prod \vee \sum)$ … DeM(k)

<u>Implication (or Conditional Exchange)(CE)</u>

k. $\prod \supset \sum$
⋮
m. $\sim\prod \vee \sum$ CE(k)

k. $\sim\prod \vee \sum$
⋮
m. $\prod \supset \sum$ CE(k)

Replacement

k. … $\prod \supset \sum$ …
⋮
m. … $\sim\prod \vee \sum$ … CE(k)

k. … $\sim\prod \vee \sum$ …
⋮
m. … $\prod \supset \sum$ … CE(k)

Exportation(Exp) / Importation(Imp)

k. $(\prod_1 \cdot \prod_2) \supset \prod_3$
⋮
m. $\prod_1 \supset (\prod_2 \cdot \prod_3)$ Exp(k)

k. $\prod_1 \supset (\prod_2 \cdot \prod_3)$
⋮
m. $(\prod_1 \cdot \prod_2) \supset \prod_3$ Imp(k)

Replacement

k. ... $(\prod_1 \cdot \prod_2) \supset \prod_3$ ...
⋮
m. ... $\prod_1 \supset (\prod_2 \cdot \prod_3)$ ... Exp(k)

k. ... $\prod_1 \supset (\prod_2 \cdot \prod_3)$ ...
⋮
m. ... $(\prod_1 \cdot \prod_2) \supset \prod_3$ ... Imp(k)

Equivalence or Biconditional(Equiv)

k. $\prod \equiv \sum$
⋮
m. $(\prod \supset \sum) \cdot (\sum \supset \prod)$ Equiv(k)

k. $(\prod \supset \sum) \cdot (\sum \supset \prod)$
⋮
m. $\prod \equiv \sum$ Equiv(k)

Replacement

k. ... $\prod \equiv \sum$ ...
⋮
m. ... $(\prod \supset \sum) \cdot (\sum \supset \prod)$ ... Equiv(k)

k. ... $(\prod \supset \sum) \cdot (\sum \supset \prod)$ ...
⋮

m. ... $\prod \equiv \sum$ ... Equiv(k)

≪≫Conditional Proof

We make available within PROPnd* the Conditional-Proof (CP) method. The rule for Conditional Proof is applied on a number of consecutive proof lines, starting at the top with the line on which an assumed premise (AP) is introduced and ending inclusive of the last line included in the proof-segment that is constructed under the assumed premise. The line justified by the CP rule is an implicational formula with the assumed premise as antecedent and the last line of the CP segment as consequent. A variety of notational schemata are used for conditional proof transactions. The critical point is that the application of the CP rule discharges the assumed premise that was posited within the proof segment. Unlike premises, which are considered self-justifying, assumed premises incur a debt that has to be paid off for the discharging of the assumed premise. The CP rule discharges the assumed premise. Intuitive justification of the mechanism is readily available. If ψ can be proven within the natural deduction system under some provisional assumption φ, then the implicational formula $\varphi \supset \psi$ is justified. Having used the assumed premise to this effect, it is no longer to be used after the application of the CP rule which discharges it. Standard Propositional Logic is characterized by the Deduction Theorem, by which proving ψ from φ is both a necessary and a sufficient condition for proving the implicational formula $\varphi \supset \psi$.

We also need to attend to management of what transactions can be carried out within the subsegment of the proof under an assumed premise. Any line that has been justified already can be repeated within the segment, and to this effect we use the rule Reiteration (also called Repetition.) Incurring additional assumed premises within a subsegment requires special attention. This is permitted but the premises must be discharged first, before the initial assumed premise is discharged. Indeed, the assumed premise that was last posited must be discharged first before the rest, and ultimately the initial posited assumed premise is discharged. Otherwise, the proof cannot be terminated. Discharges happen only by means of the application of the CP-rule, as defined.

Conditional Proof (CP)

k. ↱$\prod$
⋮
m. ↳$\sum$
m+1. $\prod \supset \sum$ CP(k-m+1)

Multi-segmented Conditional Proof (CP-...-CP)

k. ↱$^1\prod_1$ AP1
⋮
l. ↱$^n\prod_n$ APn
⋮
m. ↳$^n\sum_n$
n. $\prod_n \supset \sum_n$ CP(l-m)
⋮
p. ↳$^n\sum_1$
q. $\prod_1 \supset \sum_1$ CP(k-t)

Insertion of lines within the CP subproof segment require the rule Reiteration (Reit). This rule is to be applied also for subsegments contained within other CP subsegments. This schema is also available, for the same reason, within the IP (Indirect Proof) method which is presented below. This rule is usually omitted from textbook presentations. Its significance consists in regulating the management of available resources; it would be an error to use an assumption after it has been discharged – and this is an example of how important it is to keep track of the subsegments that are generated by means of positing assumed premises which need to be discharged appropriately, by means of the right rule, and also in the correct order. Bowing to tradition we might liberalize application of this rule, making possible its omission. It is interesting, theoretically, that this rule is not a connectives rule! Remarkably, all the rules we are using are actually connectives rules – including the CP rule, for the iplication connective, and the subsequent IP rule, for the negation connective.

Conditional Proof (CP) and Reiteration (Reit)

b. Φ
⋮
↱ k. $\prod$
⋮
m. Φ Reit(b)
⋮
n. $\sum$
↳⋮
n+1. $\prod \supset \sum$ CP(k-m+1)

We liberalize the notation, which could be encumbered with various stipulated notational details. Often boxes are used around the subproof, with smaller boxes within larger boxes for other contained subproofs. Sometimes indented lines are used to mark the boundaries of the subproofs. Or, we could mark the incurred assumed premises by a minus sign next to the labels and then place a zero superscript upon application of the CP rule and discharge of each assumed premise. Such provisions are metalinguistic as they are to be carried out by means of the justification lines of the proof. The controlling issue is that we need to abide by a fundamental structural consideration – justified as above – by which we must keep track of the debts we incur – the assumptions we posit – so that we make sure that we discharge them in due order and so that the proof can be terminated. We superimpose the metalinguistic symbols "↱" and "↳" to indicate initiation and termination of CP subproof segments. Multiple subproofs compel corresponding superscripts to the boundary-markers – "$↱^{n}$" and "$↳^{n}$". If some other device is used for demarcation – like indented lines – the superscripts are not needed.

If any assumed premise is not discharged, the proof is not considered completed; we have failed in proving the given conclusion from the premises (if we have any) and/or from the assumed premise or premises.

Multi-segmented Conditional Proof (CP-...-CP) with Subproof Boundaries

↱¹ k. $\prod_1$	AP1
⋮	
↱ⁿ l. $\prod_n$	APn
⋮	
m. $\sum_n$	
⋮	
n. $\sum_1$	
⋮	
↳ⁿ s. $\prod_n \supset \sum_n$	CP(l-m)
⋮	
↳¹ t. $\prod_1 \supset \sum_1$	CP(k-t)

Availability of the CP proof procedure (and of the Indirect Proof method, discussed below) allows us to prove no-premises conclusions, which are, semantically interpreted, logical truths or tautologies of Propositional Logic (understood as formulas that are proof-theoretically provable even without premises and, thus, from the empty set of premises symbolized by "∅".) CP can be used to prove implicational formulas that are (semantically) tautologies. The Indirect Proof (IP), shown below, is also recruited to this task.

⋚⋛Indirect Proof

The Indirect Proof (IP) rule also discharges a posited or assumed premise. Such a premise is posited expressly for initiation of the IP proof process (AP-IP) – as distinguished from positing an assumed premise for Conditional Proof. This proof-theoretic procedure is familiar from the informal proofs of Mathematics, where it has been known by the Latin name *reductio ad absurdum*. The intuitive justification is this: if we can derive a contradiction from a posited premise, then this premise must be rejected (which is accomplished proof-theoretically by introduction of the negation sign or tilde.) If the posited premise is negated, then we introduce an additional negation sign after we derive a contradiction: double negated formulas can be succeeded by lines with the same formulas without the double negation (justified by the replacement rule of Double Negation we have in our system.) Thus, Indirect Proof (aptly also called Proof by Contradiction) can be deployed to prove a positive (non-negated) as well as a negative (negated) formula. Obviously, this is another powerful tool we have at our disposal for proof of no-premises theses (which are semantically interpreted as logical truths.) There is no objection to using the method to show that a given formula is a counter-thesis (semantically interpreted as a logical contradiction or logical falsehood), by adapting the IP method so that it terminates after reaching a contradiction. This, however, is not standard practice in the bibliography.

There are different schematic options available for setting up the IP procedure. A contradiction is understood as obtaining if and only if there are at least two separate lines, one of which records some formula φ while some other line records $\sim \varphi$. This should suffice for declaring the IP subproof terminated. Or, we could mandate application of the one-directional rule schema we called Conjunction to compel formation of the line $\varphi \cdot \sim \varphi$ and then terminate the subproof. One option is to consider that, at this point, we have introduction of an absurdity symbol (which we would have to make available to our symbolic notation, symbolized as "$\perp$"). The next step is the one that introduces the negation-symbol. We could think of that step as also eliminating the absurdity symbol (but there are theoretical objections to the propriety of thinking this way – an advanced subject we cannot discuss here.) The introduction of the negation symbol we will mark as application of the IP-rule which discharges the posited AP-IP (assued premise for indirect proof.) As before, unless the assumed premise is discharged the proof cannot be considered terminated. The IP method is, indeed, uncannily powerful to the extent that it makes proofs generally easier. In the broader realm of Logic and Mathematics, this method permits proving claims about such elusive subjects as series that cannot be constructed – because of its infinitarian character.

It is rejected, in its classical version, by a school of Mathematics known as Intuitionism because of its inherently non-constructivist character. This philosophical approach to Mathematics stipulates that every proof procedure should be applied over objects that are in-principle constructible. The IP method bypasses this restriction. As such, it captures a classical understanding of mathematical proof-construction by which we can stipulate any abstract object insofar as no contradiction can be derived from this stipulation.

<u>Indirect Proof (IP)</u>

↱k. ∏	AP(IP)
⋮	
m. ∑	
⋮	
↳n. ~ ∑	...
n+1. ~ ∏	IP(k-n)

↱k. ~ ∏	AP(IP)
⋮	
m. ∑	
⋮	
↳n. ~ ∑	...
n+1. ~ ~ ∏	IP(k-n)
n+2. ∏	DN(n+2)

0.3.1 EXAMPLES

♧ Here is an example of a proof. We could actually extrapolate from the sequent we prove below and add another rule, an appropriate name for which would be "Destructive Dilemma." Nevertheless, we have drawn the line as to which rules exactly are included in PROPnd*. The PROPnd* system will be further enhanced with two procedural methods for proving, which are to be called Conditional Proof and Indirect Proof. Once this has been done, it will become obvious that many of the rules we have included in PROPnd* (or, rather, the corresponding sequents that give rise to the rule schemata of PROPnd*) can be proven by using other rules (sequents.)

$p \vee \sim p, p \supset \sim q, \sim p \supset \sim q \vdash \sim q$

1. $p \vee \sim p$	P1
2. $p \supset \sim q$	P2
3. $\sim p \supset \sim q$	P3 /.. $\sim q$
4. $\sim q \vee \sim q$	CD(1, 2, 3)
5. $\sim q$	Comm∨(5)

In the metalinguistic arrangements we make for presentation of sequents, we may also allow for the symbol "/.." to be introducing the conclusion to the right of the last premise and/or assumed premises line which open the proof sequence.

♧

$\emptyset \vdash p \supset (q \supset p)$

↱¹1. p	AP1
↱²2. q	AP2
↳² 3. p	Reit(1)
↳¹ 4. $q \supset p$	CP(2-3)
5. $p \supset (q \supset p)$	CP(1-4)

For implicative formulas that are presented as no-premises conclusions, there appears to be something of a recipe that is available for constructing the CP proof. We posit as assumed premises antecedents of the target conclusion-formula in reverse order and we then collect the assumed premises (discharging them) by applying the CP-rule.

♧ $\emptyset \vdash \sim (p \cdot \sim p)$

↱1. $p \cdot \sim p$	AP1/IP
2. p	S(1)
↳3. $\sim p$	S(1)

4. $\sim (p \cdot \sim p)$ IP(1-3)

0.3.2 EXERCISES

(1) Identify the errors in the following pseudo-applications of our formal system's derivation rules. What would correct application compel, if it is at all possible to apply the rule? You might also need to apply other rules to maneuver toward correct application of the suggested rule. A trap there is: one should not use the rule that is assumed not to be in the system; this, egregious as it sounds, can happen inadvertently.

1. $\sim (p \supset (q \cdot r))$
2. $\sim ((p \cdot q) \supset r)$ Import(1)

1. $\sim (p \vee q)$
2. $\sim p \vee \sim q$ DeM(1)

1. $(p \cdot q) \vee (p \cdot r)$
2. $p \vee (q \cdot r)$ Distr(1)

1. $(p \supset (q \supset r)) \supset \sim t$
2. $\sim t$
3. $\sim (p \supset (q \supset r))$ MT(1, 2)

1. $(p \equiv \sim q) \supset (q \vee \sim r)$
2. $\sim (q \vee \sim r) \supset (p \equiv q)$
3. $(p \equiv \sim q) \supset (p \equiv q)$ HS(1, 2)

1. $\sim (p \cdot q) \vee \sim \sim p$
2. $(p \cdot q) \supset p$ CE(1)

1. $\sim (p \equiv (q \equiv \sim r))$
2. $\sim (p \equiv ((q \supset r) \cdot (r \supset q)))$ Equiv(1)

1. $(\sim p \vee \sim q) \vee \sim r$
2. $p \vee q$
3. $\sim r$ DS(1, 2)

1. $\sim (\sim p \cdot \sim q)$
2. $(p \supset \sim q) \supset \sim (\sim p \cdot \sim q)$
3. $\sim (p \supset \sim q)$ MT(1, 2)

1. $\sim \sim \sim p$
2. $\sim \sim p$ DN(1)

(2) Complete the proofs. This might require supplying the metalinguistic justifications we write in the margin, or proof lines themselves, or both. The fact that we have redundant rules in our proof-theoretic system – for the sake of conveniencing the process and facilitating easier proofs – is evident in that we can prove some of our rules from other rules. Thus, some of our rules are not independent. Certain proofs given below show that we don't need all the rules since we can extract them from the other ones that are available in the system.

1. $\sim (p \cdot (q \vee r))$	P1
2. q	P2 $/\therefore \sim p$
3. $\sim p \vee \sim (q \vee r)$	
4. $\sim (q \vee r) \vee \sim p$	
5. $(q \vee r) \supset \sim p$	
6. $q \vee r$	
7. $\sim p$	

1. $(p \cdot q) \supset r$	P $/\therefore p \supset (q \supset r)$
2. ↱^{1}p	AP1/CP
3. ↱^{2}q	AP2/CP
4. $p \cdot q$	
5. r	
6. ↳$^{2}q \supset r$	
7. ↳$^{1}p \supset (q \supset r)$	

1. $(p \cdot q) \vee (p \cdot r)$	P $/\therefore p \cdot (q \vee r)$
2. ↱$^{1}p \cdot q$	AP1
3. ↱$^{2}p \cdot r$	AP2
4.	S(2)
5.	S(3)
6.	S(2)
7.	S(3)
8.	Add(6)
9.	Add(7)
10.	Conj(4, 8)
11. ↳^{1,2}	Conj(5, 9)
12.	CD(1, 2, 3, 4, 5, 6, 7, 8, 9, 10, 11)

1.	$p \supset q$	P1
2.	$\sim q$	P2 /.. $\sim p$
3.	↱p	AP/IP
4.	↳	
5.		

1.	$(p \vee q) \vee r$	P /.. $p \vee (q \vee r)$
2.	↱$\sim (p \vee (q \vee r))$	AP/IP
3.	$\sim p \cdot \sim (q \vee r)$	
4.	$\sim (q \vee r)$	
5.	$\sim q \cdot \sim r$	
6.	$\sim r$	
7.	$p \vee q$	
8.	$\sim p$	
9.	q	
10.	↳$\sim q$	
11.	$\sim \sim (p \vee (q \vee r))$	
12.	$p \vee (q \vee r)$	

(3) Construct PROPnd' proofs for the following given sequents. When turnstiles point in both directions, we are dealing with equivalences of the standard propositional logic. The left-facing turnstile reverses direction, indicating that what is to the right is the set of premises and what is to the left is the conclusion of the sequent. In that case, provide proofs in both directions.

a. $(p \supset q) \supset p \vdash p$
b. $(p \equiv q) \equiv r \dashv\vdash p \equiv (q \equiv r)$
c. $\sim (\sim (p \equiv q) \equiv r) \dashv\vdash \sim (p \equiv \sim (q \equiv r))$
d. $\sim p \supset \sim q, q \equiv (r \vee t), \sim q \supset \sim t, t \vdash \sim (p \cdot r) \supset (p \cdot t)$
e. $(p \vee q) \vee r, \sim r, \sim q \vdash p \vee (q \cdot r)$
f. $p \equiv q, \sim p \vee \sim q \vdash p \supset t$

g. $(p \cdot q) \vee (p \cdot t), (p \vee \sim r) \supset (r \supset \sim q), \sim r \supset r \vdash \sim q$
h. $\sim p \supset (p \supset q), \sim p \equiv \sim t, t \vee (w \cdot s), (p \supset q) \supset p$
$\vdash (\sim w \supset t) \cdot (\sim s \supset q)$
i. $\sim (w \vee (\sim q \supset \sim t)), \sim p \supset (\sim q \supset r), \sim r \vdash (w \cdot p) \vee (w \cdot$
$q)$
j. $\sim \sim p \vee \sim q, (q \supset \sim p) \supset (w \supset \sim s), s \cdot (q \vee r), t \supset w$
$\vdash t \supset s$
k. $p \supset (p \supset q), q \supset p \vdash \sim (\sim p \equiv q)$
l. $p \vee (w \cdot t), \sim w, (p \vee w) \supset q \vdash q$

(4) We can generate inconsistency proofs for given sets of formulas by showing how to derive a contradiction instantiating the form $\varphi \cdot \sim \varphi$ from the given formulas. We may stipulate schematically that this is also accomplished by having one line with φ and another line with $\sim \varphi$ in the proof. Interestingly, we can also form the conjunction of all the given formulas and generate *a proof for the negation of that conjunctive formula with no premises* – as we do for proving empty premises sequents: in this respect, we may negate the given empty-premises conclusion and apply indirect proof. Moreover, another approach we can take is the following: we pick anyone of the given formulas; we negate it and we generate a proof that this negated formula can be proven from all the others used as premises. This is because the negation of a contradiction is a thesis. In the proofs given below, showing inconsistency, fill in the uncompleted lines. Notice the liberalization of the rules permitted by laying down a thesis of propositional logic as self-justifying line – in this case the thesis laid down is an instance of the form $\sim (\varphi \cdot \sim \varphi)$.

$\nvdash p \supset q, p \supset \sim q, p$

1. $(p \supset q) \cdot (p \supset \sim q) \cdot p$ Conjunction of the Formulas
2. $p \supset q$
3. S(1)

4. p
5. q
6. ~ q

$\nvdash$ p ⊃ q, ~ p ⊃ q, ~ q

We show that any of the formulas, negated, can be proven from the rest as premises.

1.	p ⊃ q	P1
2.	~ p ⊃ q	P2/.. ~ ~ q
3.	~ p ∨ q	
4.	~ ~ p ∨ q	
5.	p ∨ q	
6.	(~ p ∨ q) · (p ∨ q)	
7.	q ∨ (~ p · p)	
8.	q ∨ (p · ~ p)	
9.	~ (p · ~ p)	Thesis
10.	q	
11.	~ ~ q	

We could also generate the following proof to show inconstancy.

$\nvdash$ p ⊃ q, ~ p ⊃ q, ~ q

1.	~ p ⊃ q	P1
2.	~ q	P2 /.. ~ (p ⊃ q)
3.		MT(1, 2)
4.		DN(3)
5.		Conj(4, 2)
6.		DeM(5)
7.		DN(6)
8.	~ (p ⊃ q)	CE(7)

Demonstrate inconsistency of the given sets of formulas by applying the PROPnd' system.

a. $\{p \supset q, r \supset s, p \vee r, \sim q, \sim s\}$
b. $\{(p \supset q) \supset p, p \equiv t, \sim t \vee q, \sim q\}$
c. $\{\sim (p \supset (w \vee (s \cdot t))), \sim w \cdot q, \sim t\}$
d. $\{(p \cdot q) \supset (w \equiv q), w \cdot q, \sim w\}$
e. $\{\sim (p \supset (q \supset p)), \sim q \equiv (r \vee (s \supset \sim t))\}$
f. $\{p \vee (q \supset r), p \supset w, \sim w, \sim r, \sim w \supset q\}$
g. $\{\sim p \equiv \sim (p \equiv q), \sim q\}$

0.4 SEMANTIC TREE SYSTEMS FOR PROPOSITIONAL LOGIC: Ʈ AND Ʈ$_{||}$

We construct a formal language for a semantic tree decision procedure, which we call "Ʈ". We use the symbols of PROP for the sake of convenience but it is understood that this is a separate formal language. This is a mechanical procedure, like the truth table, which does not require ingenuity on the part of the applying agent and is provably guaranteed to terminate (in the case of Propositional Logic) and yield the correct decisional results for determining logical properties like validity of argument forms, logical status of propositional formulas and consistency of sets of propositional formulas. This is not the case for predicate logic but it is provable that, in Propositional Logic implementations, we do not run into infinitarian constructs. Trees always terminate. We will examine first a negative tree decision procedure (so called because its validity test proceeds by adding to the premises the negation of the purported conclusion); subsequently, we will construct a positive-tree language, Ʈ$_{||}$, also called paired- or matched-trees which decide validity by constructing separate trees for the premises-set and for the conclusion and by examining whether certain systematic relations between the two (parallel) trees obtain or not.

The tree is a diagrammatic abstract object that is made up of nodes; from each node one or two branches may emanate. To produce branches application of a connective rule is required. Some connectives rules yield one vertical branch while other connective rules generate splitting or branching (two) branches. There is no structural objection to having more than two branches emanate from a given branch (which can be the case when disjunctions with more than two disjuncts are given.) The important structural restriction is that every subsequent application of a connective rule will generate branches from every branch that is still open (in a sense of 'open" which we will explicate.)

There is a distinguished node, called root of the tree, which is placed at the top (unlike what the case is with trees in the actual world, which have their roots at the bottom.) The branches are drawn downwards, which also tells us that this is basically an upside down tree if we want to think of this in relation to natural trees. Of course, this is not a natural object. A branch can be vertical or splitting. Connected branches, read downwards or upwards, constitute paths. Rather than give a rigorous definition of a path, for our present purposes we define a path as the collection of branches that are connected by means of the arrows induced by application of rules. One branch may belong to more than one paths. Evidently, a branch emanating from the root in a tree that continues between the first branching will be included in more than one paths. A branch that terminates immediately is, obviously, a one-branch path. Termination of the proof procedure settles definitively what branches and what paths the tree contains. A closed branch necessitates that the path to which this branch belongs is also closed. An open terminal branch means that the path to which the branch belongs is also open. The status of being closed, and of being open, may be applied both to branches and to paths. Caution is needed because an open branch may belong to a path that eventually closes: thus, there is no matching between closed/open branches and paths in a tree.

The nodes are labeled. Well-formed formulas of our formal language label the nodes. In practice, the formulas and the node indicators are usually omitted. We don't see the nodes themselves in the diagrammatic configuration of the tree but we consider the nodes to be there anyway. We may allow ourselves to add more conventions about what to write around the tree. For instance, we may use positive integers to the left of the tree to keep track of the lines formed across by the formulas on nodes. To the right of the tree we may opt for writing justifications of how we get the formulas that are on the nodes. These are metalinguistic symbolic superimpositions. We will see such details later but let us first consider a generalized diagram of this interesting abstract object we call tree.

⊙	**Root (Distinguished Node)**
⇓	**Vertical Branching**
⊙	**Node**
⇙ ⇘	**Splitting Branching**
⊙ ⊙	**2 Nodes**
⋮ ⋮	
⊙ ⊙	**2 Nodes**
⊕ ⊗	**[Termination – open/closed paths]**

A branch closes if and only if its path contains an individual variable letter and the negation of that letter. We may stipulate that the path (thus, the branch) closes immediately when this occurs. A closed branch closes the path passing through it, which has the individual letter and its negation on it. Such a path represents a logically impossible state of affairs: an assignment of truth values to the individual letters of the given formula or formula, which is logically impossible. The parallel to this in the case of the truth table is a row across which all the formulas receive valuations of false. We read the truth values differently in the case of the tree: the negated individual letters represent those letters as false and the non-negated or positive individual letters represent those letters as true in the state of affairs. Every path represents a state of affairs – with closed paths, as we saw, representing logically impossible or absurd states of affairs. Open paths, on the other hand, are characterized by valuations – assignments of truth values to the individual letters – which satisfy the given formula, or all the given formulas. In a truth table, this is a row across which all the formulas receive valuations of true (hence, are consistent.) This already suggests that a test for consistency of a given set of formulas, by the tree method, consists in establishing that there is at least one open path in the terminated tree of the given formulas.

We will first learn how to construct a tree for one or more formulas of our language L. Let us consider first the simple case of a tree for one well-formed formula of L: if the formula is ϕ, then we can call its tree "$\text{T}(\phi)$." [First time we baptize some item we use quotation marks; but subsequently we don't need them.] We can also construct the tree of more than one formulas: $\text{T}(\phi_1, \ldots \phi_n)$. The formula or formulas about which we construct the tree are inserted into the root and are considered to be the <u>initial list</u> of the tree. The trees we will deal with in Propositional Logic are not infinite – they are always guaranteed to terminate – but other logics can have infinite or non-terminating trees. A node of a tree is terminal if it is labeled by either an individual letter or an individual letter with the tilde in front of it and no more rules can be applied to induce a branching underneath it.

To learn how to construct Ʈ(ϕ), we will consider first special trees which we use to build formula trees. Those special trees are called connectives trees. They are the building blocks of trees for formulas. Each connective and each negated connective has its own special connective tree – or connective rule. The only exception is the connective we know as negation: *there is no tree for negation* – or, as we will say, there is no tilde tree. In addition to the connectives rules, implementing instantiations of the schemata for the connectives rules, we need structural rules that lay out the framework within which we proceed.

0.4.1 STRUCTURAL RULES FOR Ʈ

The wff or wffs that are given establish the root of the tree. In the special case of argument validity tests, the conclusion of the tested argument form is added to the tree's root *negated*.

In addition to the branching rules there are termination rules: these rules specify under what conditions a branch of the tree is closed and under what conditions a branch remains open or, as we say, is satisfied. The node as well as the path on which the closed branch belongs are also considered closed; the same applies in the case of open branches/nodes/paths.

The tree itself is considered closed if and only if all its paths are closed. Since every path is composed of one of more branches arranged vertically and from top to bottom, a closed tree will have both all its paths and all the branches of its paths closed. It is provable as a theorem in Metalogic that every tree we can draw for argument forms in our logical system terminates after a finite number of steps. Such steps are carried out by applying the specified rules. Terminations are to nodes which have atomic letters or singly negated atomic letters from {p, q, ...}.

◊ Opening Rule (OR): Write out the given wff or wffs argument form above the top node (root) of the tree. If the tree check is for validity of argument form, the premises are written vertically, one under another, and the putative conclusion is written to the right of the last premise separated by the symbol "/..".

◊◊ In the case of Argument Validity Tests, a First Step is mandated: This is called the Negated Conclusion Step (NC): Attach the negated conclusion of the logical form underneath the premises of the logical form; the negated conclusion is to be treated as a provisional additional wff at the root of the tree.

◊◊◊ Termination Rule (TR): When no more application of rules is possible (when we are down to only atomic letters and negated atomic letters for the nodes of a path in the tree), then the path is considered terminated: we can also say that the final branch of the path is terminated. A terminated branch can be open or satisfied (and marked underneath by "⊕") if and only if it is not closed.

A path is closed (or unsatisfied, or contradictory) if and only if along this path we find both an atomic letter (any letter) and the negation of the same letter. We legislate that the branch closes immediately when this occurs. We mark closure underneath by the symbol "⊗". This is a closure rule (CR) for the branch. Obviously, a branch is open or satisfied if and only if it is not closed.

A tree is terminated if and only if all its paths are terminated. A tree is closed if and only if all its terminated paths are closed.

As we will see, the connectives rules are decomposition trees, which means that they show the rules for how to break up formulas. We will have connectives trees for all the connectives (except for the tilde) and for all negated connectives (in which case tilde-tilde is not excluded.) Those rules or partial trees will be sufficient for construction of the trees of any formulas or sets of formulas. Let us recall that our set of connectives is: {~, · , ∨, ⊃, ≡}

Accordingly we will need the following rules: ·R, ∨R, ⊃R, ≡R, ~~R, ~·R, ~∨R, ~⊃R, ~≡R.

0.4.2 CONNECTIVES RULES FOR T

Vertical Rules: ~~R, ·R, ~ ∨R, ~⊃R

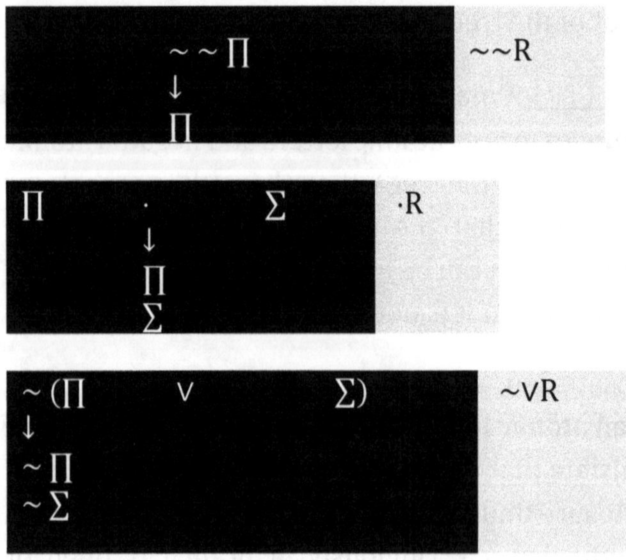

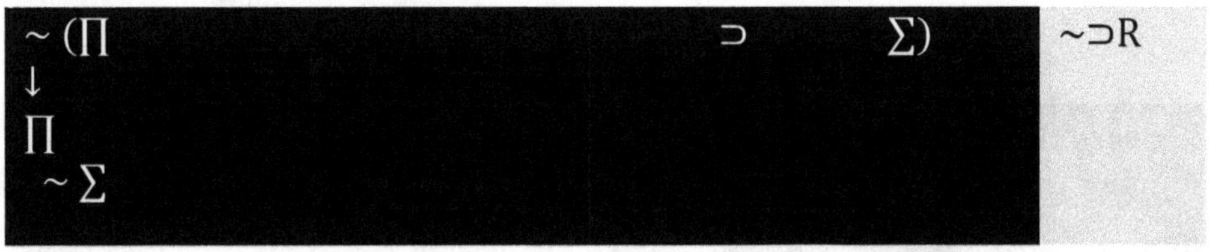

Branching Rules: ∨R, ⊃R, ~·R, ≡R, ~≡R

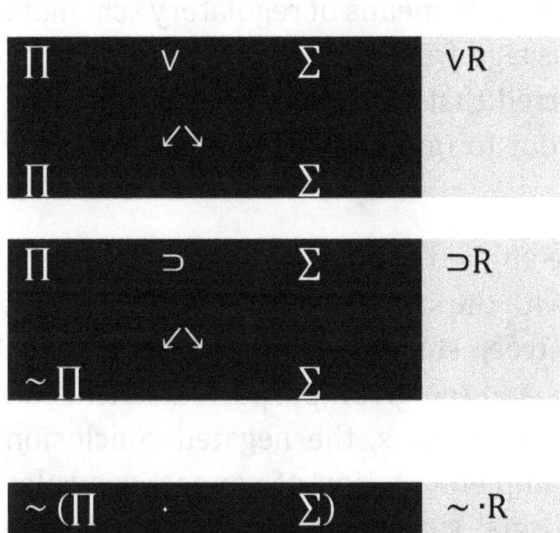

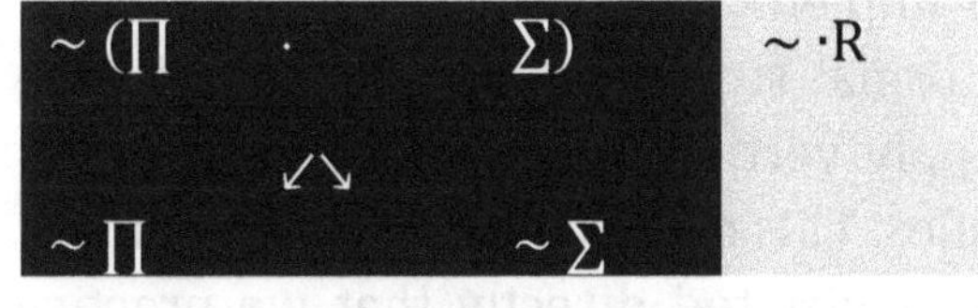

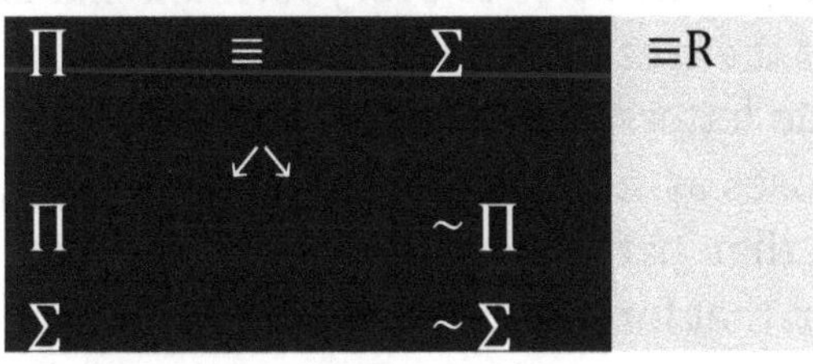

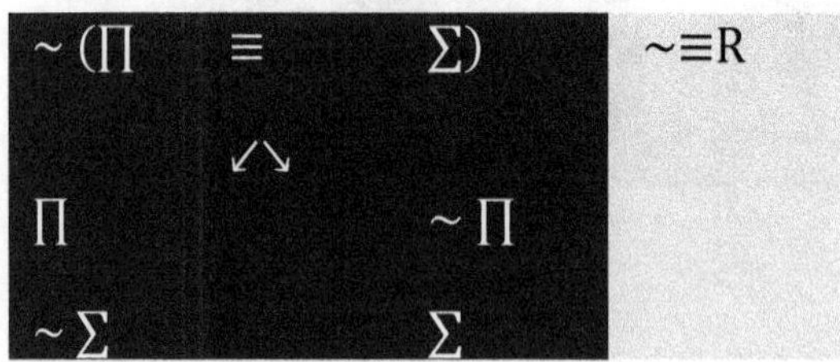

Since the connectives rules are given by means of regulatory schemata (recipes for how to manage passage from one node to the next by means of branching), it is expected that we know how to substitute into tokens of those rules in order to implement them. We will give examples subsequently.

The Ţ decision procedure can be applied to check for validity of given argument forms. In accordance with the structural rules of the method, we establish the root of the tree, starting from the top, which comprises the premise formulas and the given purported conclusion formula with the tilde in front of it (thus, the negated conclusion formula is placed in the root.) Implementation of connectives rules follows. As a matter of extra-formal strategy – not dictated by the structural rules – we should apply vertical rules before we proceed with application of branching rules. This makes the tree less expansive in its dimensions. We have not legislated directly that we practice <u>saturation</u>: that means that we apply the rules to every formula that is composite – for the purposes of the tree formalism counting as non-composite or simple only atomic letters and negated atomic letters. Because we allow that a path closes as soon as we find any letter and the negation of that letter up that path, we cannot also mandate saturation. It is implied, however, that for non-closed paths, saturation is requested; this follows from the termination rules which require termination to consist in having only atomic letters and negated atomic letters on all final nodes - which are then declared to be terminal nodes.

The validity test is this: The given argument form is valid if and only if there is no open path (or, equivalently, all paths are closed, or all terminal branches are closed, or the tree is closed.) The justification is easy to discern. Closure of the tree means that there are no logically possible assignments of truth values to the atomic letters, for which the root formulas are all true: of course, the root contains the premises as true and the conclusion as false (since the conclusion has been placed in the root negated.) Thus, the closed tree for validity test means that there is no logically possible state or valuation for which all the premises are true and the conclusion is false. If any (at least one, one or more) path is open upon termination (if the tree is not closed but open), then the given argument form is determined to be invalid. There is at least one logical option or logically possible state or valuation (assignment of truth values to the atomic letters) for which all the premises are true and the conclusion is false. This is, indeed, a counterexample to the given argument form - defined as the truth values to the atomic letter variables for which all the premises are true and the conclusion is false. Thus, we can read counterexamples to given argument forms from the open paths of the trees.

We can also use the $\mathcal{T}$ system to check for consistency of a given set of propositional formulas. We construct the tree of the formulas (the tree with root comprised of all the given formulas) until termination. The given set is consistent if and only if there is at least one open path of the tree. Such a path furnishes the truth value assignments to the atomic variable letters for which all the given premises (the root formulas) are true. As explained earlier, valuations are read from scanning the atomic letters, so that non-negated letters are true and negated letters are false.

Determination of the logical status of a given propositional formula is not always a straightforward affair. We construct the tree for the formula. The given formula is a logical contradiction (logical falsehood, invalidity) if and only if its tree is closed. The justification is straightforward. There is no logically possible assignment of truth values to the atomic components, for which the formula is true. If the tree of the given formula is open, however, we do not have a direct access to determining its status. Consider the individual atomic letter, which may be considered as a degenerate case of a semantic tree. All its paths are (vacuously) open or, we can say, the tree is open. This is not, however, a tautology. Thus, we cannot depend on observations about the open paths to determine status of tautologousness. The consideration we apply is this: A formula is a tautology if and only if its negation is a contradiction. This is handy because we have already seen how we may establish status of contradictoriness. Accordingly, to check whether a given formula is a tautology, we place its negation in the root of the tree we construct: the given formula is a tautology if and only if the tree is closed. Finally, the status of logical contingency (indefiniteness, indeterminacy) is established by constructing the trees for the given formula and for its negation: the formula is a logical contingency if and only if both trees (the formula's tree and the negated formula's tree) are open.

To implement the tree method with a view to determining what relations obtain between any given two formulas, we may first form a formula conjoining the given formulas: to establish logical equivalence we conjoin the formulas by means of the equivalence sign. To check whether the first implies the second we conjoin them by means of an implicative formula with the first as antecedent and the second as consequent – and conversely to check if the second implies the first. Having thus formed a new formula, we have to check if this new formula is a tautology: only in that case do we show that the relationship obtains.

0.4.3 POSITIVE TREE METHOD: $\mathcal{T}_{||}$

We construct $\mathcal{T}_{||}$ as the system for implementation of decision procedures that fall under the broader category of <u>positive trees</u>. It is unusual for this type of system to be included in a text. The term betrays the fact that validity testing by means of this method is not carried out by negating the conclusion formula. - with the term suggested by the fact that validity checks do not depend on negating the conclusion. Other names for this type of semantic tree method are "paired" and "parallel" semantic trees.

We retain the same symbolic resources and grammatical conventions as in $\mathcal{T}$. The connectives rules are the same as in $\mathcal{T}$. The structural rules are altered in the case of the validity check - but not for implementation of checks for consistency. For logical status, in the case of tautology-check, the parallel-trees method can be adapted to a test that is not available for the negative trees method. Contradictoriness is established in the same way as in $\mathcal{T}$. We can think of $\mathcal{T}$ and $\mathcal{T}_{||}$ as overlapping systems or, perhaps more elegantly, as branches of the same system with variously adapted structural rules for each case.

The positive trees method requires, structurally, two trees for checking validity of a given argument form. One tree has the premises as its root while the other tree, placed parallel to the premises tree, has its root labeled by the conclusion of the given argument form. The whole diagrammatic object is now the ordered pair of the two trees, with the premises tree to the left of the conclusion tree:

$$\mathcal{T}_{||} = < \mathcal{T}_{||pr}, \mathcal{T}_{||c}>$$

The structural rules being the same as in $\mathcal{T}$, we complete the two trees to termination. At that point we test for validity. We are interested only in the open paths of both trees. We revert to a metalanguage to form the so-called premise sets and conclusion sets: those sets have as members exactly the atomic letters and negated atomic letters for any open path. Thus, we may have to form multiple premise and conclusion sets. It is crucial that any open path can be continued with splitting branches underneath its final node – so that one branch has some letter (not on the path) and the other branch has the same letter negated. We do not have to enforce this branching-to-saturation if the validity condition – to be given – is satisfied; otherwise, we have to continue to saturation in this fashion. Only the premises tree is amenable to saturation. One final term we need: a set X is covered by another set Y if and only X is a superset of Y (which means that every member of Y is in X.)

The validity test, stated summarily, is as follows: The given argument form is valid if and only if every premise set is covered by some conclusion set.

Certain observations are in order: It is not necessary that every conclusion set is involved in the covering tally. But every premise set should be involved – and be shown to be covered. The determinative account is taken after saturation is carried out in the manner we indicated – unless it is not needed because, even without saturation, it is established that every premise set is covered by some conclusion set. Clearly, more than one premise sets can be covered by the same conclusion set – there is no regulation to the contrary. It should be noted also that the covering does not require that the premise set is a proper superset of some conclusion set – only that it is a superset of some conclusion set. (A set X is a proper superset of set Y if and only if X is a superset of Y, as defined above, and there is at least one member of X not in Y.) It is obvious that a condition requesting proper supersethood must fail: in the case of the trivially valid argument form with an atomic letter as premise and the same letter as conclusion, we have automatic supersethood but not proper supersethood. It is interesting to notice also that nothing would change if we kept saturating such a tree with additional letters. This comment is not particularly helpful here but it will be relevant when we turn to adaptation of this method for predicate logic.

Our supply of metalinguistic symbols superimposed on the trees requires additional resources: while retaining the symbols for open and closed paths and other such signs, we need dedicated symbols to signifying whether a premise path is covered (which means that the premise set we read off that open path is covered): we use the symbol "◉" for that purpose whereas we use "▬" to signify that the open premise path is not covered after termination and saturation. To indicate provisional termination we may use "⊕": this means that we are left with no formulas on any nodes of the tree, which compel application of some connective rule; and we have as provisionally terminal nodes (labeled by) individual letters or negated individual letters. We mark this as a "provisional" termination because saturation may allow us to add more branches. By saturation we mean branching so that for some atomic letter ⌜ p ⌝ one added branch is labeled by this letter and the other branch is labeled by the negated letter, ⌜ ~ p ⌝: the branches should be in splitting mode.

Implementation of the system for consistency is the same as in Ţ. In that case there is no parallel conclusion tree. We might wonder why we need to talk of the positive tree method beyond validity; the reason is that there is a way to check tautologousness by using parallel trees. In this case, the premises tree has its root labeled by the empty set symbol, "∅". It is crucial to note that this is not the same as a closed path. For the empty-set premises root, we may, and ought to, continue with saturation. A closed path, on the other hand, cannot be developed by saturation. The given formula labels the root of the conclusion tree. The tautologousness test is based on the well-known semantic fact that a tautology, and only a tautology, is validly entailed by the empty premises set (thus, it renders, as conclusion, any argument form valid.)

For the purposes of the tautology test, we branch for saturation; we may or may need to branch for all the individual letters in the conclusion form. We establish tautologousness by applying our validity test for positive trees: every premise set is covered by (is a superset of) some conclusion set. To establish status of contradictoriness, we negate the given formula and place it at the root of the conclusion set. For contingency check, we need to form the parallel trees both for the formula and for its negation as conclusion-trees with the empty set as premises trees. We provide an example.

Establish whether is ⌜ q · ~ q ⌝ a logical contradiction.

∅ ?⊢T|| ~ (q · ~ q)

Saturation
∅ => {q}, {~ q}

Premises Sets
P1 = {q}
P2 = {~ q}

Conclusion Sets
C1 = {~ q}
C2 = {q}

Covering
P1 ⊇ C2
P2 ⊇ C1

We need an ad hoc adjustment to cover the case of vacuous validity. An inconsistent set of premises validly entails any conclusion in the case of formal systems within the standard Propositional Logic. To accommodate this, notice we do not resort to arrangements made for the empty set of premises which is a different matter as we indicated above. We simply need to stipulate that a closed premises tree is covered vacuously. This is shown in the following example.

p · ~ p ⊢$_{TII}$ (p ∨ q) ⊃ (p · q)

1. p · ~ p		1. (p ∨ q) ⊃ (p · r)
↓	<<>>	↙ ↘
2. p		2. ~ (p ∨ q) p · r
3. ~ p		↓ ↓
⊗		3. ~ p p
		4. ~ q r
		⊕ ⊕

No Premise Sets => Vacuous Validity

The case needs to be considered also, in which the conclusion tree is closed. This corresponds to the case in which the conclusion formula of the proffered argument form has the logical form of a contradiction. We stipulate that a closed conclusion tree cannot offer any covering. We note, however, that the case of the vacuous validity escapes this restriction: in the case of vacuous validity, there is no need for covering to be actuated – the covering is considered to be effected vacuously. Thus, we can say that a closed conclusion tree does not cover any non-closed premises tree.

0.4.4 EXAMPLES

p ⊃ (q ⊃ r) T$_{II}$ ⊢ (p ⊃ q) ⊃ (p ⊃ r)

1. p ⊃ (q ⊃ r)		1. (p ⊃ q) ⊃ (p ⊃ r)
↙ ↘	<<>>	↙ ↘
2. ~ p q ⊃ r		2. ~ (p ⊃ q) p ⊃ r
◉ ↙↘	↓	↙ ↘
4. ~ q r		3. p ~ p r
◉		4. ~ q ⊕ ⊕
		⊕
↙↘		

5. **p** **~ p**

◉ **◉**

Covering:

P1 = {~ p} ⊇ C2 = {~ p}

P2 = {~ q} – not covered

Saturation => {p}, {~ p}

P3 = {p, ~ q} ⊇ C1 = {p, ~ q}

P4 = {~ q , ~ p} ⊇ C2 = {~ p}

P5 = {r} ⊇ C3 = {r}

Compare the proof by using the negative variant of the semantic truth method, which we implement by Ʈ.

p ⊃ (q ⊃ r) Ʈ ⊢ (p ⊃ q) ⊃ (p ⊃ r)

1. p ⊃ (q ⊃ r)				
2. ~ ((p ⊃ q) ⊃ (p ⊃ r))			**Negated Conclusion**	
↓				
3. p ⊃ q				**~⊃(2)**
4. ~ (p ⊃ r)				**~⊃(2)**
↓				
5. p				**~⊃(4)**
6. ~ r				**~⊃(4)**
↙	**↘**			
7a. ~ p	**7b. q**			**⊃(3)**
⊗	**↙ ↘**			
		8b. ~ p	**8c. q ⊃ r**	**⊃(1)**
		⊗	**↙ ↘**	
			9c. ~ q r	**⊃(8c)**
			⊗ ⊗	

For our next example we implement Ʈ||.

∅ Ʈ|| ⊢ p ⊃ (q ⊃ p)

```
1.     ∅                               1. p ⊃ (q ⊃ p)
      ↙  ↘               <<>>               ↙  ↘
2. ~ p    p                            2. ~ p   q ⊃ p
   ◉    ↙  ↘                              ⊕    ↙  ↘
3.     q   ~ q                         3.      ~ q   p
       ◉    ◉                                   ⊕    ⊕
```

P1 = {~ p}
P2 = {p, q}
P3 = {p, ~ q}
C1 = {~ p}
C2 = {~ q}
C3 = {p} P1 ⊇ C1 P2 ⊇ C3 P3 ⊇ C2 and P3 ⊇ C3

0.4.5 EXERCISES

(1) Answer the following questions:

a. How can we read counterexample values (the truth values assignments for the atomic letters, for which all the premises are true and the conclusion is false) by using the T method to check for validity? What about the case of the $T_{||}$ method?
b. Can we given definitions of validity, invalidity, consistency, inconsistency by appealing to the Negated Semantic Tree Method? What about the case of the Paired Trees Method?
c. Why can't we declare a formula to be a tautology if all its paths are open upon termination?
d. Validity of an argument form can be defined as inconsistency of the set comprised of all its premises and the negated conclusion. Can we make sense of this by referring to the Semantic Tree tests for validity/invalidity and consistency/inconsistency?
e. What do we mean by saying that the closed tree of premises does not generate an empty set of premises in the case of the

Paired Trees method? Does this distinction between empty sets of premises and closed tree of premises matter in the case of the Negative Tree Method? Explain.

(2) Deploy the Ʈ and Ʈ$_{||}$ methods to make semantic decisions for the given formulas below. If it is one formula only, the check should be for logical status (tautology, contradiction, or contingency); if it is a set of formulas that is given, then the test should be for consistency/inconsistency; and if it s a set of formulas with a distinguished formula as propounded conclusion, then the check should be for validity/invalidity. Reflect that the Consistency/Inconsistency check is the same for the Negative and Positive Tree methods. Turnstiles pointing in both directions require examination in both directions. Notice that requests for consistency checks do not use the turnstile: the formulas to the right of the turnstile ought to be understood as being joined by means of inclusive disjunction, not by conjunction as the case ought to be for examining consistency.

a. $\vdash ?\ ((p \vee q) \cdot (p \supset r)) \supset ((q \supset r) \supset r)$
b. $p \supset q \supset r\ ?\dashv\vdash ?\ (p \cdot q) \supset r$
c. $(p \supset q) \supset p \vdash ?\ p$
d. $(p \supset q) \cdot (p \supset r)\ ?\dashv\vdash ?\ p \supset r$
e. $\vdash ?\ (p \vee (p \supset (q \equiv \sim p))) \supset \sim (p \supset (q \vee \sim p))$
f. $\vdash ?\ \sim (p \vee q) \equiv (\sim p \cdot \sim q)$
g. $(p \supset q) \supset p\ ?\dashv\vdash ?\ p \vee (q \cdot p)$
h. $\sim (p \vee \sim \sim (q \supset \sim t)) \vdash ?\ (p \supset (q \supset w)) \supset \sim w$
i. $p \equiv (q \supset \sim t)\ ?\dashv\vdash ?\ (\sim p \cdot \sim q) \equiv t$
j. $\sim w \equiv (q \vee \sim p) \vdash ?\ w \supset ((p \cdot q) \supset \sim w)$
k. $\{\sim p \cdot (q \supset (\sim t \supset w)), \sim w \vee \sim q, t \supset (w \supset q)\}$
l. $\{(p \cdot q) \equiv (\sim w \cdot \sim t), p \supset w, q \supset \sim t, (w \supset t) \supset p\}$
m. $\{\sim ((p \supset q) \supset \sim w), w \cdot (t \vee \sim q), t \supset (\sim w \supset \sim (q \vee t))\}$
n. $p \supset (q \cdot t) \vdash ?\ (p \vee r) \supset ((q \cdot r) \supset (p \cdot (q \vee r)))$

o. $\sim (p \supset (p \supset (p \supset q)))$?⊣⊢? $\sim (p \cdot (p \supset (q \supset p)))$
p. ⊢? $(p \equiv (p \equiv (q \vee p))) \vee (p \equiv (p \supset (q \vee p)))$

0.5 TRANSLATIONS FROM ENGLISH INTO THE PROPOSITIONAL LOGIC IDIOM PROP

The propositional language PROP will be used to translate propositions expressed by meaningful (rightly assertable, if true) sentences of English. The status of a proposition is a vexed matter but we don't need to enter into philosophic animadversions. Some authors use "statement" or even "sentence" instead of "proposition" because of the metaphysically loaded associations of the latter term. All we need to bear in mind for our current purposes is that an indefinite number of sentences – written as well as uttered, but also visual and other communicative acts – can express the same meaning; not to mention that we can can have translations of such sentences into other languages or expressed by other means of communication – and they continue to express the same meaning. Any meaning can be expressed by an indefinite, open-ended number of appropriate means. What is expressed in all such cases, what all the sentence and other means have in common, is what we have named "proposition." We have tokens or instances of the proposition in each case but we do not worry about the metaphysical status of such entities. We use capital letters, possibly with superscripts from the denumerable set of positive integers, to effect the translations. Because the subsequent predicate logic idiom will be extending the propositional PROP language – retaining all the connective symbols, as defined in PROP – we need to cover the subject of translations into PROP before we proceed to our study of predicate logic.

All the connectives of PROP are truth-functional. They are defined to have as meanings the conditions under which they are true (and, symmetrically, also by the conditions under which they are false.) This shows us that logical meaning is a matter of truth value in Propositional Logic. Because logical meaning is, more broadly, a matter of reference in the so-called Extensional logical languages, we may also draw as an inference that the connectives are defined as referring to their truth conditions. The computational and decisional effectiveness of Propositional Logic is a happy result of this but there are also limitations, when it comes to translating, which stem from this simple basis for the selection of connectives. Non-truth-functional expressions of language cannot be translated into PROP and, insofar as this is due to the truth-functional character of the connectives, they cannot be translated into the predicate idiom either. There is a view that predicate languages can capture meaningful talk about such matters as time, for instance, by quantifying over moments or temporal periods (with quantification symbols made available in predicate logic.) Nevertheless, the expressive power of non-extensional logics, like Modal Logics, cannot be reached from within the standard logics (propositional and predicate.)

Non-truth-functional expressions of language are all those that cannot yield a unique determination as to true or false; instead, context-variability (any kind of context that is dynamically shifting) affects whether the expression is true or false when its ultimate atomic components (individual sentences) are given as true or false. For instance, "it is necessarily true that a triangle has three angles" is itself true but a shift in context (which can be broadly characterized by replacing the sentence on which "necessarily" applies) may result in false even if the new sentence is true. Thus, "it is necessarily true that the planet Earth has one moon" is true in the actual context but it is not true in all contexts: we may conceive logically of a context – an alternative, non-actual but logically possible state of affairs – in which the Earth has more or fewer than one moon. Thus, "necessarily" is not a truth-functional component: when applied to atomic component propositions that are valued as true or false, it does not yield always, across all conceivable contexts, the same truth value as output. What counts as atomic proposition is itself dependent on the observations we have made about truth-functionality. For instance, any proposition expressed by a sentence "necessarily, __" is simple if the line placeholder is occupied by a sentence that expresses a truth-functionally simple proposition. To trace this successfully we must begin at the other end and develop the ability to recognize truth-functional compound expressions. All such expressions, by definition, can be rendered as having in them truth-functional connectives; the propositions that cannot be further analyzed after the truth-functional components have been identified are to be considered for our purposes as individual or simple or atomic.

It is not possible to provide an exhaustive classification of truth-functional expressions of language; the reason for this is the complexity and organically evolving dynamism of a natural language like English. We provide a token list below. Our notational idiom PROP has a small number of truth-functional connectives in it. We show how other expressions, which could be captured by some truth-functional connective not in our formal language, are to be rendered by using only the connectives we have available in PROP. Intuitions desert us when we ascend beyond the binary degree. We should also note that, perhaps charitably to our efforts, other monadic truth-functional connectives, which are mathematically definable, besides negation, do not correspond to linguistic expressions. The same is true of certain mathematically definable binary connectives as well - like, for instance, one that yields the constant output value false for all possible combinations of input values.

The criterion as to whether a linguistic logic-word is truth-functional or not has been given in terms of compositionality - we should remember this: a logic-word is truth-functional if and only if availability of the truth values of its atomic components suffices for determination of the truth value of the whole proposition - with atomic components being decidable on the basis of recognizing truth-functional words and carving the compound sentence accordingly. For example:

(S1) *The Earth is a planet and our moon is its satellite.*
The verb to be is logically inert in its role as copula; this is significant. As identity between two objects, it cannot be captured truth-functionally in Propositional Logic. The logic-word "and" is immediately recognizable. This word is usually truth-functional – but beware of cases in which its meaning is "after" or "before" because such expressions are not truth-functional. In such cases, we regard the whole proposition as simple or atomic (if no other truth-functional logic-words are present.) Although the second conjunct is properly expressed by "the Earth's moon is its satellite," we realize that we have a conjunction – thus, a compound sentence. No other truth-functional connectives – or non-truth-functional words – are present. Although grammatically this is a simple sentence, its logical grammar compels us to recognize it as compound. It has the logical form of a conjunction. Its logical meaning is truth-if-and-only-if-both-conjuncts-are-true.
(S2) *He took his shoes off and entered the house.*
In this case "and" means "before" – or "after" if read with what follows "and" first. Therefore, it is not truth-functional. True propositions can be combined with "after" or "before" so that in some cases the result is a true proposition but in other cases it is not true. (Similarly, false propositions can be combined to variable effects.) Because "true before true" is not either true or false determinately – and the same for other valuations of the apparent parts of the sentence – we do not have a truth-functional connective. Thus, the proposition expressed by (S2) scans in Propositional Logic as simple, not as compound.
We give now cues to truth-functional translations. In passing, without space to enter into this intricate subject, we note that especially the implicative connective of our logic is inadequate in translating "if-then" and related particles of English. Our implicative connective or conditional (called material conditional) does not capture the logical behavior of inference or deducibility in proofs, or entailment. It can be thought of rather as an inclusive disjunction so that "if p, then q" has

the meaning of "either not-p or q" and also of "not-(p and not-q)." If we use this conditional in translation, we mean to capture the minimalistic meaning of "if-then" understood to consist in "not possibly p and not-q." There is a view that a natural language like English relies on usage-conventions so that if-then and other recalcitrant expressions are not presented if simpler, less misleading, more relevant linguistic formations can be offered instead: because of this, it is argued, the problems we encounter with the truth-functional biconditional are obviated to a great extent.

We may consider our translation successful if what is lost does not have an impact on assessing the logical properties we investigate – validity, consistency, tautologousness, contradictoriness, contingency, and logical relations that may obtain between propositions. This directive implies that we should always strive for a translation that reveals more of the logical form of the proposition. For instance, (S3) has more than one logical forms we could capture by translating. It should be self-evident which one reveals more; straightforwardly, that is the logical form that has the largest number of connectives in it. (At one extreme, the atomic proposition form ⌜ p ⌝ is a logical form of any proposition. But it reveals very little indeed. It is not surprising that a proposition has more than one logical forms

Neither A nor B:	$\sim A \cdot \sim B$
Only if A, B:	$B \supset A$
If A, then B:	$A \supset B$
B if A:	$A \supset B$
A unless B:	$\sim (A \cdot B)$, $\sim (A \equiv B)$, $\sim B \supset A$, $A \vee B$

The case of "unless" is instructive. It is usually given as an inclusive disjunction, which can be justified by appealing to the semantic fact that, in the standard Propositional Logic, ⌜ $p \vee q$ ⌝ is logically equivalent to ⌜ $\sim q \supset p$ ⌝, "if not-q, then p" which seems a reasonable translation of "unless." But inclusive disjuncts can both be true when the whole compound is true and this sounds wrong for the case of "unless." Now, it is true that the truth tables we can draw for all the formulas (or, more accurately, for the logical forms of the formulas) shown above as translating "A unless B" agree for certain pairs of truth-value inputs. The fact that our formal language cannot calibrate fine-grained nuances is due to the limited number of its expressive resources. As we have pointed out, we pay a price, when it comes to translations, as a trade-off for the decisional simplicity and efficiency of this logic. When it comes to translating an expression like "unless", we may exercise judgment in each case. We recall the overarching imperative that assessment of logical properties should not be impacted by our translation.

A if and only if B:	$A \equiv B$
Either A or B (or both):	$A \vee B$
Either A or B (and not both):	$(A \vee B) \cdot \sim (A \cdot B)$, $\sim (A \equiv B)$, $(A \cdot \sim B) \vee (\sim A \cdot B)$
$A \supset B$	Given A, B. B follows from A. It must be B if A. It cannot be A and not-B. On condition that A, B. But for B, A. [Thus, B captures the necessary condition for A to be true: consider also: If not-B, then not-A.] If A, then B.

<u>Sentences with Non-Truth-Functional Monadic Logic-Words</u>

It is necessary that A.
It is unlikely that A.
It is likely that A.
It is possible that A.
It is probable that A.
It is morally obligatory that A.
It permissible that A.
It is known that A.
It is believed that A.

It is physically impossible that A.
It is a law of nature than A.
It is proven that A.
It is disproven that A.
It is true specifically at x that A.
It is actually true that A.
It will be true in the future that A.
It was true in the past that A.
It has always been true that A.
It will always be true that A.
It is computable that A.
It is desirable that A.

---- Sentences with Non-Truth-Functional Dyadic Logic-Words
A is true before B is true.
A is true after B is true.
It is provably A if it is B.
It is true that A simultaneously that B is true.

Propositional Connectives cannot express Vagueness
It is rather true that A.
It is sort of true that A.
It is more true than it is false that A.
For example:
It is sort of true that John is tall.
We take the proposition, as expressed by the whole sentence, to be true or false. We have no semantic means for expressing "sort of true" or "sort of false" or any other semantic degree. It is highly controversial that we should ever allow for degrees of truth.
There are philosophically and linguistically motivated views that we should withhold true and false from certain meanings expressed by declarative sentences: for instance, when existential presuppositions fail, misleadingly, as when "John's children are sick" when John has no children. A more radical view assigns a third truth value, neither true nor false, to the meanings of such sentences – in this way treating them

as propositions of a different type. Similar cases are made for nonsensical propositions that are used in language in practice, or for predictions about the future, or for propositions whose classical truth value (true or false) cannot be determined. Such views motivate construction of truth-functional languages with more than two truth values. We do not examine such logics here and, clearly, we do not have available such unorthodox symbolic resources for our translations.

We do not have Predicate Logic symbols
The following have to be translated as single propositions.
All students are hard-working.
Some students are hard-working.
John is hard-working.
Everyone likes someone or other.
There is someone who is liked by everyone.
Everyone is liked by someone.
Someone likes someone else.

The following have to be translated as two single propositions that are conjoined by the truth-functional connective we find in the sentence.
John and Mary are students. [John is a student and Mary is a student.]
If Mary is a student, then John is a student.
Either Mary is a student or John is not a student.
Mary and John are friends.
[This is interesting. Is the meaning of this "John is Mary's friend and Mary is John's friend?"]
The propositions we translate into PROP are understood to have all their context-dependent and variable or dynamically affected elements fixed. Our target propositions do not change truth value over time or by being matched to any other shifting context. This means that we take our propositions to be containing all references to the dynamic contexts that affect them. The proposition expressed by "Socrates sits" could be thought of as being true when Socrates sits and false

otherwise (vacuously false when Socrates does not exist) but the proposition "Socrates sits at such-and-such-a-time" is either true or false definitely. As an example, we may also consider the proposition expressed by the sentence "Caesar crossed the Rubicon." There are indexical elements in this content: the time, the place, to say the least. The proposition we are translating is to be considered as the one expressed by "Caesar crosses the Rubicon at such-and-such-a-time, ---, ___, _-_, ..." where all the line-type symbols stand for specifications of indexical factors - like space, time and anything else that can be context-dependent and its shift could impact the truth value of the proposition.

0.5.1 EXAMPLES

1. Translating an argument.
Only if no one studies, does everyone fail. But someone or other always studies. Therefore, it is not the case that everyone fails.
KEY for Translation:
No one studies: N;
Everyone fails: E;
Someone or other studies: S.

1. $E \supset N$
2. S /.. $\sim E$

Here we have a translation of an argument whose semantic characteristics make valid. Nevertheless, our translation yields an argument form that checks as invalid, as we can attest by applying one of the decision procedures we have learned (truth table or semantic tree, for instance.) This is a good example of how inadequate expressive resources may lead to inaccurate results in determining logical properties. This is not an internal defect on the part of the formalism itself: if the translation reveals sufficient logical structure, then the formal language generates a faithful translation that checks correctly. The problem is relative shortage of expressive resources, as we have said; a genuine defect of the formal system would consist in failure to decide correctly even if the translation were successful with

respect to showing the crucial logical structure of the target proposition.
Incidentally, we could add a premise to the above argument form, which we can justify by appealing to logical relationships beyond Propositional Logic, so that our propositional resources can accommodate the translation so that the argument form we obtain checks as valid. Although this is rarely covered in texts, we may essentially correct – either by omitting or by adding defensible propositions. If we were using the truth table as our decision procedure, we would be adding a formula and constructing the truth table for the totality of formulas we attained. In some other cases, we can correct the truth table by deleting rows – as when, for instance, we have interpretations for which certain truth value outcomes cannot possibly obtain. These are extra-formal intrusions into the system, of course, but it is a testament to the decisional power and versatility of our apparatus that such impositions can be accommodated to good effect.

Adding to the preceding argument:

3. If someone or other studies, then it is not the case that no one studies.

[We could have added the stronger claim "someone or other studies if and only if it is not the case that no one studies." The proposition expressed by the statement is stronger than the one we added in the sense that it implies it without being implied by it. It suffices, however, to add the weaker claim because we still obtain a valid argument, as we should. It seems that our imperative about getting logical properties right, combined with a parsimony rule, dictates adding the weaker claim.]

1. $E \supset N$
2. S
3. $S \supset \sim N$ /.. $\sim E$

This is a valid argument form.

2. *If it Monday, then, if the weather is not rainy, the game is not cancelled. But the game is cancelled only if the weather is rainy and it*

is not Monday. Unless it is Monday and the game is cancelled, the weather is rainy if and only if the game is cancelled only on Mondays.

[We may want to check for consistency of this collection of claims.]

KEY:
It is Monday: M.
The weather is rainy: R.
The game is cancelled: C.
1. M ⊃(~ R ⊃ ~ C)
2. C ⊃ (~ R · ~ M)
3. ((M · C) ∨ (R ≡ (C ⊃ M))) · ~ ((M · C) ∨ (R ≡ (C ⊃ M)))

[We symbolized "unless" by means of an exclusive disjunction. It is interesting to check if we have the same results, pertaining to consistency, when we symbolize as an inclusive disjunction and also when we symbolize as negated conjunction (which is the relation of contrariety.)]

3. *No existential proposition should be provable deductively from premises all of which are either definitions or logical principles. But, if existence can be treated as a logical property, there is at least one such proof of an existential proposition. Therefore, some existential propositions may be provable from only definitions and logical principles.*

[Words like "necessarily" and "should" are not really doing any work here insofar as we pack in the content of the propositions the information that these are deductive proofs. Similarly, even though the expression "it is provable that__" cannot be symbolized truth-functionally, this does not affect reaching the correct decision result when we check for validity. This argument is not valid - more accurately, its argument form is not valid. The antecedent of the second premise, "if existence can be treated as a logical property" is thereafter omitted in the conclusion. The argument should not be valid and its truth-functional rendering checks the argument form as, indeed, invalid. Once again, as in the previous example, we take

advantage, extra-formally, of the logical relationship between "some" and "none" – with the one being the negation of the other.]
KEY:
Some existential proposition is deductively provable from premises all of which are either definitions or logical principles: P.
Existence can be treated as a logical property: E.

1. ~ P
2. E ⊃ P/.. P

The premises prove, if anything, that existence is not properly treatable as a logical property. Of course, it should always be kept in mind that an assault against a deductive argument moves to regarding soundness after the validity test has been conducted successfully: if the argument form is valid, the argument that has that form can still fail if at least one of its premises is not to be warranted – in which case the argument is valid but unsound.
If one were to add to the premises of the above argument that existence is a logical property – would this be warranted? – then we can prove from the premises logical absurdity (we can deduce both that some proposition is and is not properly provable from premises which are definitions or logical truths.

0.5.2 EXERCISES

(1) Make a key and translate into PRED the propositions expressed by the following English sentences.

a. John and Jill perform a clown act together.
b. Unless Jill performs, John doesn't.
c. Neither John nor Jill are clowns.
d. If Jill is a clown, then if John is clown they perform together.
e. Not only does John perform as a clown, but Jill does too.
f. John is a student although he is also a clown.
g. It is not possible for both John and Jill to be good students.

h. It is preferable that John and Jill perform together rather than separately. [* can you use the same symbols you used for "John performs as a clown" and "Jill performs as a clown?"]
i. John performs after Jill performs.
j. John performed yesterday and Jill will perform tomorrow. [* can you use the same symbols you used for "John performs" and "Jill performs?"]
k. Jill performs as a clown and sings and dances.
l. Only if John performs as a clown does Jill perform as a clown too.
m. It is not known whether John is a student but Jill is.
n. Insofar as a Jill performs as a clown, she performs together with John.

(2) Given the key, render the symbolized propositions in properly idiomatic English.

KEY: A: The aliens are among us.
H: The aliens have hostile intentions.
T: The aliens have advanced technology.
I: The aliens are interested in studying us.
K: The aliens conduct systematic kidnappings.
G: The government is keeping the alien presence a secret.
P: The aliens have visited in the past.
E: The aliens have educated the human species.

a. $H \supset (T \supset I)$
b. $\sim ((E \cdot (T \supset G)) \cdot (H \cdot K))$
c. $(E \supset T) \supset (I \supset K)$
d. $\sim I \cdot \sim H$
e. $G \equiv (P \cdot (K \vee H))$
f. $(K \cdot G) \supset (H \cdot \sim E)$

$(H \cdot K \cdot G) \supset (\sim E \vee (E \supset I))$

I. PREDICATE LOGIC

The next step up from Propositional Logic (also called Sentential Logic or Statement Logic) is Predicate Logic (also called First-Order Logic or Quantification Theory.)

We deploy a formal symbolic idiom we call PL. We enrich this with polyadic (many-place, n-ary) predicate letters as well as with the identity symbol. The passing beyond monadic predicate logic to relational predicate logic and relational predicate logic with identity has profound metalogical consequences affecting various logical properties but, in the present text, we are not engaged in the study of such subjects.

In any formal language that is an idiom of Propositional Logic we have as ultimate carriers of logical meaning individual propositions - also called atomic propositions or atoms. If there are no truth-functional connectives in a proposition, we consider it as an individual or atomic proposition regardless of its apparent grammatical form. For instance, the following sentences of English express atomic propositions.

S1: We have to stay after class today because of a fire drill. ("Because" and "have to" are not truth-functional parts of language.)

S2: It is possible that it is Monday today. ("Possible" is not truth-functional.)

S3: It is Monday today.

S4: John is a student.

S5: John loves Mary.

Another way of thinking about an atomic proposition is this: we can disregard what it talks about and focus only on the fact about this proposition that it is either true or false and it cannot be both true and it cannot be neither true nor false. Hence, disregarding the linguistic content of the proposition, we have to use a capital letter to symbolize it - per our specified grammar - and we have to use a small letter from {p, q, ..., p_j, ...} (where j is a positive integer) to symbolize the logical form of this proposition.

The propositions expressed by S4 and S5 above have interesting grammatical structure, which we cannot translate into Propositional Logic. We do not have the expressive resources we need for this in Propositional Logic. Thus, we cannot express either names like "John" and "Mary" or predication of attributes like "__ is a student" and "__loves---" of the entities picked out by those names. In other words, Propositional Logic cannot capture the inner structure of attribute-predication; nor can it express quantificational phrases like "everyone" or "at least one person." Propositional Logic treats atomic or individual propositions as lacking logically interesting internal structure or as being unanalyzable. This has serious consequences and it counts as a significant limitation of Propositional Logic. Some students of logic will not even contemplate Propositional Logic as adequate for any purpose although this has a deeper historical source: that Propositional Logic is inadequate for expressing the truths of Mathematics - and it was this purpose, related to Mathematics, which motivated development of modern logical techniques. Of course, deductive logic is not about the non-logical content of propositions - it is not about what propositions talk about. The problems associated with Propositional Logic must be presented in relation to what deductive reasoning has as its subject - for instance, the study of argument forms and the determination as to whether such forms are valid or invalid. To see how Propositional Logic is limited consider the following argument made in English, which ought to be valid on the basis of the standard intuitions of users of the language but does not check as valid when translated into Propositional Logic.

Caesar crossed the Rubicon.

/.. Therefore, there is at least one person who crossed the Rubicon.

Both the premise and the conclusion are individual or atomic propositions. The argument can be symbolized in some idiom of Propositional Logic as follows:

C /.. S

Its logical form is:

p /.. q

Of course, this is an invalid argument in Propositional Logic. There is a counterexample to this argument form: simply assign true to the premise and false to the conclusion. And yet, our informal assessment, sanctioned by the competence we have as users of the language, compels us to take this as a valid argument. If someone named Caesar crossed the Rubicon, then definitely there is at least one person (the one named Caesar) who crossed the Rubicon. There is no way of making the premise true and the conclusion false; and yet, in Propositional Logic, there is such a way.

This is a puzzling situation but, having read the opening of this section, you can now figure out what has gone wrong. The argument is valid indeed but assessment of it as valid requires that its internal structure of predication is expressed and that the quantificational phrase ("there is at least one such and such") is also translated into the formal language. Propositional Logic has no resources for capturing these elements of internal analysis of the proposition. In contrast, the logic we will now introduce, Predicate Logic, is equipped with such symbolic resources and has all the needed adjustment in its formal machinery for checking an argument like the one above as valid. While Propositional Logic deals exclusively with unanalyzed propositions (or takes propositions as unanalyzed to deal with them), Predicate Logic has what is needed to handle analyzed propositions (or to take propositions as analyzed and treat them accordingly.)
Before we continue with laying out the details for the study of Predicate Logic, we may address a question which may be raised in this respect: Could it be that Predicate Logic also suffers from a similar predicament relative to some other logical language? In other words, could it be that arguments which, assessed informally but competently, come across as valid are evaluated as invalid by Predicate Logic because it lacks the resources needed? Indeed, here is an example of such an argument:

Caesar crossed the Rubicon.

/.. Therefore, it is possible that someone crossed the Rubicon.

Predicate Logic - like Propositional Logic to which it is attached - cannot express non-truth-functional expressions like "it is possible that___." Hence, Predicate Logic too will check the above argument as invalid, although it is quite obvious that this is a valid argument. To gain resources and formal instruments that allow us to capture the form of this argument and assess it as valid we need to ascend to what is called Modal Logic. Such kinds of logics - Modal Logics, as a family of systems - are controversial but they are in fact studied extensively and have copious applications in practice. We will not enter into the philosophic controversy about modal logics here.

♛The Grammar of Predicate Logic PL

We need a name for the formal language through which we will approach the task of studying standard predicate logic. We call it "PL." From now on, we will not need parentheses around PL. (It is like saying for someone that her name is "Mary." After the initial statement of "her name is 'Mary'," when we talk about Mary we don't place quotation marks around the name anymore.)

PL keeps PROP (the basic propositional language) intact and extends it. What this means is that the logical connectives are the same as in standard Propositional Logic. After all, PROP is to be regarded as an idiom of standard Propositional Logic. Of course, since the connectives are the same, they are still defined over the truth values set of L, which is

VALUES = {T, F}.

We keep all the symbolic resources for propositional variables as we have them in L, using the same conventions for letters as we specified in the grammar of PROP. Of course, since we now have access to resources that allow us to symbolize the internal structural elements of propositions, it is questionable why we would ever want to represent any propositions as unanalyzed. Be that as it may, we retain the option, whether we choose to exercise it or not, and consider ourselves to be retaining PROP and to be building on top of it by introducing new symbols and by specifying the grammar for these symbols. A formal grammar, as we recall, consists in the precise specification of rules for concatenating the symbols of a formal language so as to form symbolic expressions or formulas that are then regarded as well-formed (or wffs, well-formed formulas.) Of an expression that is well-formed we say that it belongs to the set of WFFs or well-formed formulas of the formal language. We can symbolize this as "$\varphi \in$ WFF" for a well-formed formula symbolized by "φ".

To ascend to PRED (our idiom for Predicate Logic) we add symbols that will prepare us for the semantic labor we will be doing. In the first place, we regard these symbols as squiggles that are to be governed by a given grammar. When we introduce the semantic approach subsequently, these symbols will have to serve certain roles. The semantic approach will have the things symbolized refer to certain objects. Before we take the semantic step, we just lay out a game with symbols - telling us what symbols we have and how we are allowed to put them together or concatenate them. This first step is called syntactical. Let us think ahead to what we will have to include in the semantic approach. This will tell us what syntactical resources we need to provide in the first place (because we will be mapping the syntactical resources on the resources of the semantic approach although we will not show details in this introductory text.) In general, think of a syntactical system as one with rules about how to play a game of putting together meaningless symbols; think of a semantical system as one that will allow us not only to manipulate our symbols correctly (as the syntactical system also does) but also to tell certain meaningful stories about the things symbolized.

The semantics of PRED will require, to begin with:
Predicates: these are logical predicates, not exactly like the grammatical predicates in ways you will soon find out.

Names: From the standpoint of the grammar for PROP, these are letters that we will call individual constants. When we introduce semantics for PROP, you will have to learn to think of these names as labels or tags that pick out objects in a specified way. You might be content thinking of names in general in this way, as simply labels that identify and track items or entities (but this is actually one of the hardest philosophic problems.) Either way, the names of PRED are just labels that name or refer to specified items.
Pronouns of a certain kind: these are pronouns that are systematically, deliberately ambiguous. As syntactic symbols, we call them individual variables. In the semantic approach, these variables will fail to refer to things and this is systematic and intended. (There are alternative ways of building a predicate logic system but we ignore that here.) You will soon find out why we need such individual variables. In the syntactic phase of our construction, the individual variables play a significant role in the building of the grammatically correct formulas (or well-formed formulas, wffs) of Predicate Logic.

Quantificational Phrases: The semantics of predicate logic will require symbolic expressions for phrases like "all such and such things" and "some such and such things." Accordingly, in the syntactical build-up we have to include symbols whose mapping (also called interpretation) will give us the quantificational phrases we mentioned above. The predicate logic we study will not have quantificational phrases matching such expression in language as "most such and such things" or "few such and such things", etc..

We should be using different symbols in the syntactical and in the semantic constructions but we simplify and stipulate are the same: this is to be regarded as a matter merely of convenience. If you will, it is like a coincidence by which two different approaches to system-building happen to have the same symbolic vocabulary.
We will also need symbols for the above kinds of items in our metalanguage - the mix of English and symbols we use to talk *about* PL. Strictly speaking, different symbols should be used in the

metalanguage from the symbols used in the PL-language (which is also called Object Language.) For the sake of simplifying our conventions, we can agree to use the same symbols.

We are ready to introduce the symbolic and grammatical conventions of PL.

SYMBOLS

All the symbols of PROP (our Propositional Logic language) are retained.

Briefly, here they are, presented in a way that is easy to grasp intuitively. "p" etc. stand for individual or atomic propositions, "φ" etc. symbolize well-formed formulas (not necessarily atomic or simple propositions), and "j" as subscript can be drawn from the set of natural numbers.

GRAMMAR(L): p_j/~ φ/φ · ψ/φ ∨ ψ/φ ⊃ ψ/φ ≡ ψ/)/(

For PROP, symbols are added as follows. In the future, more symbols may be added.

INDIVIDUAL VARIABLES:

$x, y, \ldots, x_1, x_2, \ldots, z, \ldots, w, \ldots, u, \ldots, v, \ldots$

We could have an infinite number of such variables. Starting with "x" we may continue to the end of the English alphabet, and we can also use subscripts from the set of positive integers {1, 2, 3, ...}.

INDIVIDUAL CONSTANTS or NAMES:

$a, b, c, d, e, \ldots, a_1, a_2, \ldots$

Starting with "a" we can continue all the way to when we clash with "x" and subsequent letters (since these letters have been reserved for the individual variables). We can also use subscripts from the set of positive integers and we do have available to us a denumerably infinite stock of such symbols, as with the individual variables.

PREDICATES (also called PREDICATE CONSTANTS, PREDICATE LETTERS or NON-LOGICAL CONSTANTS):

Starting from "A" we have all the capital letters of the English alphabet available to us. We can also use subscripts and superscripts, from the positive integers up to a (denumerably) infinite stock.
Next, we will explain what role is played by the superscripts, but let us also stipulate this:
If there is no ambiguity, given the context, we can omit the superscripts that indicate *arity*. Now we will learn what is meant by "arity."

Monadic or

$A^1, B^1, C^1, \ldots, A^1{}_1, B^1{}_2, \ldots, F^1{}_1, \ldots, G, \ldots,$

Predicate letters symbolizing unary predicates will be accompanied by exactly one individual variable or individual constant letter in proper grammatical formation. Intuitively, and looking forward to the semantic interpretation, an unary predicate is intended to catch the logical behavior of a linguistic attribute that can be predicated of exactly one entity: for instance, "John is a student" and with the logical predicate, unlike the grammatical predicate, indicated as follows: "John is-a-student" or "student(John)." If you want to reflect more deeply on this, notice that the verb to-be (which has been philosophically troublesome in the history of thought) is rendered inert in this way. For one thing, we don't need a symbol for "to be" and we don't treat "being" or "existing" as a logical predicate.
If there is no ambiguity, given the context, we can omit the superscripts that indicate *arity*.

Dyadic or Binary Predicates

$A, \ldots, A^2{}_1, \ldots, B, \ldots, F^2, G^2, H^2, \ldots, F^2{}_1, F^2{}_2, \ldots, R^2{}_1, \ldots$

Continuing on the preceding characterization of arity, a binary predicate letter must be followed by exactly two individual variable letters or exactly two individual constant letters for proper grammatical formation. Intuitively, such linguistic attributes are binary as "John loves Mary" with the logical, unlike the linguistic, binary predicate indicated as in "loves<John-Mary>." Recall the brief discussion of ordered pairs in an earlier section and mark that <John-Mary> is meant to be ordered; "<Mary-John>" may not have "love" predicable of it if, alas, it is not true that Mary loves John. Binary and generally n-ary predicates are relations or relational predicates. This brilliant move allows modern Predicate Logic to handle relations; this option had eluded Aristotle and other ancient logicians.

<u>Nary or n-place Predicates</u>
If there is no ambiguity, given the context, we can omit the superscripts that indicate *arity*.
Generally, <u>n-adic or n-ary Predicates</u>:

A^n_1, ..., F^n_1, G^n_1, H^n_1, ..., F^n1, F^n2, ..., R^n_j, ...

For trinary or triadic linguistic attributes consider, for example, "John is between Mary and Sally" or, indicating the logical trinary predicate, "is-between<John, Mary, Sally.>" Intuitions begin to fail us when we rise beyond trinary predicates but formal logic can continue to study such predicates with no problem.
If there is no ambiguity, given the context, we may omit the superscripts.

QUANTIFIERS
Universal Quantifier: ∀
Existential Quantifier: ∃
These are the only two available quantifier symbols in our formal language for predicate logic. Quantifier symbols must be accompanied each by an individual variable. By themselves, quantifier symbols with the accompanying variables per each quantifier symbol do not constitute well-formed formulas (wffs). They must be concatenated to the right with predicate symbols, of whatever arity, and numbers of variables corresponding to the arity. Individual constants may occur instead of individual variables. We will present details when we turn to the grammar of our formal system subsequently.
We do not have symbolic resources for expressing any other from a plethora of conceivable quantifier phrases which abound in natural language. For instance, we cannot express “most of” or “a few”. An error that must be forestalled from the beginning is that our existential

quantifier is to be understood as meaning "at least one" without specific commitment as to whether most or only a few objects are meant as having the indicated property. We are able to handle alternative quantifiers but, in so doing, we would be constructing alternative, non-standard logics. We would also be venturing into the realm of alternative logics if we allowed our models to have more than one domains and we assigned two distinct pairs of universal and existential quantifiers to be ranging over two separate domains. This is called sortation but our approach, dovetailing standard predicate logic, foreswears sortation; so, our pair of quantifiers are unsorted or are dedicated to range always over one specified domain.

$\forall x\phi$ x is not bound by any other quantifier in ϕ

$\exists x\phi$ x is not bound by any other quantifier in ϕ

IDENTITY

This is the familiar algebraic symbol "=". We will make appropriate grammatical and, subsequently, semantical stipulations about this symbol and what it stands for. But we can already plant a certain notion about identity. This is actually treated as a logical predicate. As such, it has to be a binary predicate and we could have elected to symbolize it as "I^2" which is something you may find in some older textbook. If you are wondering why this one logical predicate is distinguished for special treatment, we will have more to say subsequently. For now, think about this. If we take any chance attribute of language - for instance "John is a student" or "John loves Mary" - we can intuit immediately that the propositions of such sentences are what we have learned to consider logical contingencies. It is logically possible for such propositions to be true but it is also possible for them to be false - it doesn't matter for deductive reasoning which one of the two possible situations is the actual one in our context. But for identity regarded as an attribute what is special is this: the proposition of a sentence like "the number two is identical with itself" has to be a tautology. And this is not the case only in Mathematics or when we talk about numbers and such abstract objects. It has to be a logical truth that every entity is identical to itself. It could even be said, and has been said, that this must be the most "obvious" logical truth - although we might be weary of entrusting obviousness which sounds like a characteristic of a psychological reaction.

We will have more to say about identity but the above is a preliminary foray into what gives us grounds for making this binary logical predicate - the identity predicate - distinguished. We don't know yet how the machinery of predicate logic is adjusted to incorporate a distinguished logical predicate but we will find out in due time. As it is, keep in mind now that the predicate logic system we are building, PROP, is, more specifically, a system of predicate-logic-with-identity. We could have started with some version of PRED without identity and then add identity to get a system we could call "PL=". This makes no difference for our purposes. Having identity in the system gives expressive means for covering many more phrases of language - and of the language of Mathematics - than we would be able to capture without identity. In fact, infinitely more propositions can be expressed as analyzed propositions after identity has been included but you will need to wait to find out how identity is used in translations from a language like English into our formal language PL.

If you are wondering, there are more things for which symbols can be added in PL: functions for one! We will not do this in the present text but it is feasible and further increases our expressive power.

Finally, if you are still wondering about existence: we do not, and should not, treat existence as a logical predicate in our setup. If we had a more elaborate or advanced text in mind, we would find out that there are ways, actually, for *predicalizing* existence - treating existence as a logical predicate - but our standard system, PROP, cannot do this. We will have an opportunity to see why. Many students of logic have regarded this as a healthy feature of the standard predicate logic. Actually, more than you can imagine now rides on this because there are some famous philosophic proofs touching on high-stakes matters, which stand or fall with allowing existence to be treated as a logical predicate. This would mean, for instance, that "God exists" or, as we shall say for the logical predicate, "exists(God)", should be considered meaningful because it is properly constructed in its logical grammar (no question about its surface grammar); but the objection has it that this is not meaningful because the logical grammar of "exists" is such that "exists" is not a logical predicate!

x = y, ...
a = b, ...

Connectives

The connectives symbols (which are also called, interestingly, logical constants symbols) are as in PROP with the set of symbols being:

$\{\sim, \cdot, \vee, \supset, \equiv\}$

The connectives are defined - semantically, when dealing with truth values - over the same set of values as in L:

$VALUES^2 = \{T, F\}$

Parentheses (which are considered to be auxiliary symbols) are used, as in L, only when they are needed for disambiguation (to remove and prevent ambiguity of expressions.)

Function Symbols

Although we can perform all the symbolic tasks that are incumbent on first-order logic formalism without using special symbolic resources for functions, we can facilitate presentations and achieve parsimony by using such symbols. Moreover, there are interesting logical-philosophic issues, into which we cannot enter here, that might militate in favor of using function symbols instead of resorting to alternative methods. The relevant cue is that the function symbols in predicate logic, if included, are treated as terms; this places them in the same category as individual variables and individual constants. Terms may have as referents, or denotata, objects of the domain of a model. Individual variables may be thought of as failing, systematically, to denote objects and, as such, they are to be compared to inherently ambiguous pronouns of a natural language (although it is doubtful if everyday languages have such pronouns or if context is supposed to settle the referent of any pronoun.) Individual constants refer or denote - or receive their valuation by referring - to objects in the domain or universe of discourse. So do function symbols. If we look into natural language, the functions of formal predicate logic are motivated by capturing the logical behavior of complex noun phrase - like "the sister of such-and-such" or "the sum of such-and-such and so-and-so." It should be clear, then, that the function symbols will have degrees: for instance, "the sister of such-and-such" is unary or one-degree because it receives one input which, incidentally, is itself an individual constant. The case of the function symbol for "sum" shows us an example of a binary or two-place function symbol since it requires two inputs (also called, rather confusingly but traditionally, arguments of the function.) The arity of a function symbol may be indicated by means of a numerical superscript, as we have allowed for predicate letters, but in this case too we may also drop the superscript. Subscripts are also allowed, as usual, up to the cardinality of the denumerable set of positive integers. Functions can serve as inputs to other functions.

$$f^1(x_k), f^1{}_j(a_m), g^1(y), \ldots, f^2(f(x)), f^2(g(a)), \ldots, f^2(x, y), f(x, a), f^2(a, b), g^2(f(x), f(y)), \ldots, f^n(x_1, \ldots, x_n)$$

I.1 PRED-GRAMMAR

We extend PROP, the formal language we built for Propositional Logic, to construct PRED. The symbols for connectives are as in the propositional idiom we used.

$$\{\sim, \cdot, \vee, \supset, \equiv\}$$

It is not always the case, but we allow for individual propositional variable letters to be incorporated in PRED, and, accordingly, for any well-formed formula of PROP, which is constructible on the basis of those resources, to be available for PRED too. One might wonder why we need such symbols once we have entered into the territory of analyzed propositions - with symbolic resources that allow representing internal parts of meaningful sentences - but we might consider translating the proposition of a sentence like "if it is believed that there are no solution, everyone is doomed." The antecedent of this proposition has in it the non-truth-functional part "believe" which, arguably, compels translation by means of an atomic individual letter; the consequent, on the other hand, can be symbolized analytically by means of predicate symbolic resources. Thus,

[p] is well-formed.

Now, we consider the rest.

$[F^1x]$, $[F_1{}^1{}_x]$, [Fx], $[F_1x]$ are well-formed (or, belong to the set of WFFs: well-formed formulas of PL.)

$[F^1a$, $[F_1{}^1a]$, [Fa], $[F_1a]$ are well-formed (or, belong to the set of WFFs: well-formed formulas of PL.)

$[F^2xy]$, $[F_1{}^2xy]$, [Fxy], $[F_1xy]$ are well-formed. (We are not stipulating that "x" and "y" cannot be the same: hence, for instance, [Fxx] is well-formed.)

$[F^2ab]$, $[F_1{}^2ab]$, [Fab], $[F_1ab]$ are well-formed. (We are not stipulating that "x" and "y" cannot be the same: hence, for instance, [Faa] is well-formed.)

$[F^nx...w]$, $[F^nx_1...x_n]$, [Fx...w], $[Fx_1...x_n]$ are well-formed.

Consider φ and ψ as (metalinguistic) symbols of well-formed formulas (wffs) of PRED with the individual variable ⌜ x ⌝ in them. (Because they are metalinguistic symbols, they are not placed within the special corner-brackets used for the *mentioned* individual-variable symbol from the object-language.) Then, the following formulas are well-formed formulas (wff) of the formal idiom, when the individual variable symbols accompanying the quantifiers are presumed not-bound in the formal phrase to which the quantifiers are applied. Not being bound means that the individual variable symbol is not within the scope of any quantifier: once the quantifiers are applied, the unbound (also called free) variables become bound. This suggests already that we disallow vacuous quantification – quantifier symbols with no variables within their scope. Individual constants do not require, or permit, binding.

$\forall x\varphi$, $\exists x\varphi$, $\forall x\exists y\varphi$, $\exists x\forall y\varphi$, $\forall x\varphi\exists y\psi$, $\exists x\varphi\forall y\psi$, ...

We retain the convention of infix notation for placement of the symbols for the logical connectives: this means that instead of writing, for instance,⌜ ⊃ p q ⌝ which is prefix notation, we write ⌜ p ⊃ q ⌝. (We use the special brackets because these are wffs from the object language of the formal idiom PROP – cited here in a metalanguage.) Notice that we abandon prefix in favor of infix notation for the placement of the ⌜ = ⌝ symbol which stands for a distinguished predicate letter. Obviously, we do this bowing to common usage both in Mathematics and in everyday contexts although we could, of course, stipulate prefix notation: "= x y"

$x = y$, $a = b$, ...

are well-formed.

$f^1(x_k)$, $f^1{}_j(a_m)$, $g^1(y)$, ..., $f^2(f(x))$, $f^2(g(a))$, ..., $f^2(x, y)$, $f(x, a)$, $f^2(a, b)$, $g^2(f(x), f(y))$, ..., $f^n(x_1, ..., x_n)$

are well-formed.

Nothing else is well-formed. (This is called the "closure clause.") Any sequence of symbols, which is not a wff (not well-formed) is to be considered as gibberish.

Before we proceed to the semantics of PROP, let us first linger around issues of grammatical correctness.
We must repeat that, strictly speaking, we should set up a separate formal language for the semantic system and then map our syntactically constructed language above onto that other (the semantic) formal language. We simplify by taking this formal language, which we have already named PROP, and importing it, as it were, into the semantical construction.

This is a splendid opportunity, in passing, for reflecting on one of the deeper - but also fundamental and in a sense preliminary - aspects of the study of formal logic. Notice that the syntactical construction above has no things to which the symbols refer. [Perhaps, in an attenuated sense, the symbols refer to the kinds of abstract things referred to by symbols constructed within formal systems. But this is a detail. The point is that we have no tales to tell to account for what the symbols refer to.]

Note specially that the symbols of a syntactical system remain *meaningless* insofar they have no specified items to which to refer. As they are for now, before the semantical interpretation is set up, they are meaningless squiggles that are concatenated in accordance with specified grammatical conventions to form correct sequences or phrases (what we have called well-formed formulas.) This is the syntactical approach to the study of formal logic. When we move to the semantical stage, we will supply meanings for the symbols.

Realize that what we just said is underpinned by a special view of meaning: on this view, the meaning of a thing X that is meaningful is some item Y to which X refers. The shorthand expression is: "meaning is reference". Or, we can say, logical meaning is referential. Consider a phrase like "the morning star." The meaning of this phrase is straightforward but its reference (the star also named "Venus") turns out to be referring to the same object to which the phrase "the evening star" refers! Two radically different connotative meanings but the same denotative meaning or reference. There is a view according to which logical meaning is such that "connotation *is* denotation" and "meaning is extensional." This is controversial but we will not linger here on this subject. This view is called Extensionalism.

The Extensionalist reduction of logical meaning to reference applies to names and predicates but also to propositions. You will see how this will all come together when we move to the semantics of PL. But we should already take care of what to make of the *logical meanings of propositions*: what is the kind of thing to which an atomic proposition p refers to so that it has (or is) a logical meaning? The answer is: the referent of a proposition is a truth value (true or false, not both and not neither.)

Thus, the meaning of a proposition is the same as its truth value.

$\surd(p) \in \{T, F\}$

This says that,

the value of a proposition belongs to the set of VALUES, {T, F}, but we can read it as also saying that the meaning of a proposition is its truth value, which is one of true or false (members of the set of values if we put this in set-theoretic language.) Perhaps, we should better say, that a proposition is a logical meaning in the sense that it has a referent - which is a truth value.

In Propositional Logic, as we have already seen, the truth value - which is the logical meaning - of a compound proposition depends functionally on the truth values of its components. This establishes an important principle, which is called Compositionality of Logical Meaning. The grammars of spoken languages like English cannot achieve this target but a formal language like that of Propositional Logic can. When we move to predicate logic, the question arises as to how this principle is going to be adjusted or if it is to be preserved at all. In Predicate Logic, unlike in Propositional, we treat propositions as analyzable: we have expressive resources - symbols - for internal parts of propositions like names, predicable attributes, and quantifying phrases. When we move to the semantics of PROP, we will see how the truth value of the proposition will be dependent on the logical meanings or values of the parts: notice, however, that we cannot say that the dependence is on the truth values of the parts because names and logical predicates do not have truth values. (They do have values, though, as you will see soon.) Because our formal grammar accepts putting together symbols to form grammatically correct expressions that may not state a proposition, we must accept that Compositionality of Meaning fails in Predicate Logic. We can offer a notion of what is afoot. Assume that "it" is a pronoun that is decidedly ambiguous: we have no way of disambiguating it or of determining what "it" refers to. For the purpose of setting a formal grammar for an idiom like PROP, we must allow a phrase like "it runs" to be grammatically correct; but this sentence – to be called open sentence – cannot possibly have logical meaning (which means it cannot be decided as to whether it is true or false, given its referential indeterminacy.) In this way, we have, unfortunately, lost strict compositionality of meaning, which we could relish in the case of Propositional Logic.

The philosophic position we have called Extensionalism insists that this way of viewing logical meaning is sufficient for the study of formal logic. When we move from Propositional Logic to Predicate Logic, we are still within the boundaries of Extensional Logic. Not so if we ever ascend to Modal Logic, but that is a different matter. Things to notice about extensionalism are the following:
1. Every meaningful expression symbolizes something that is meaningful insofar as it has an appropriate kind of thing it refers to: in the case of propositions, this means that individual proposition symbols symbolize propositions and propositions refer to truth values - hence, propositions are logically meaningful (as they ought to be) by being either true or false; (also, nothing expressed by sentences that is not true-or-false can be logically meaningful or be a proposition).
2. Things that refer to the same item must be identical - must be the same thing - regardless of what initial impressions you might have about them. Thus, any two tautologies, or any two contradictions for that matter, have the same constant referent (respectively true or false), which tells us that they have the same logical meaning.
3. The meaning of the whole or compound must be functionally dependent on the meanings of the parts. In the case of propositions, as we know by now, this means that when we are given the truth values of the component atomic propositions we can determine uniquely (functionally) the truth value of the whole or compound proposition. This is the *Principle of Compositionality of Meaning*. Notice that we couldn't make this work in the grammar of a language like English even if we stipulated that meanings are referents. But in Propositional Logic, the trick works! On the other hand, we could say that anything for which this doesn't work does not belong in the realm of formal logic.
4. Any two things that have the same meaning can replace each other within a compound without changing that compound's meaning.

Indeed, if we replace any proposition within a compound proposition by a logically equivalent proposition, then the truth value of the compound is not affected (it is still true, if it was true before the replacement, or it is still false if it was false before the replacement.) This is known as the *Law of Extensionality* or Leibniz's Law or the Law of the Substitutivity of Equivalents.

Free and Bound Variables

The crucial significance of this subject will become manifest when we take the semantic approach to the construction of PL. Approaching the subject syntactically first, we have it that the following phrases are grammatically correct or well-formed; we have placed no restrictions to the contrary.

Fx, Gy, Bz, Hx_{33}, W^2u, P_2y, Rxx, Q^2xz, Fax, Gyb, H^1c

Given our grammatical instructions, we can pick any capital letter, with or without subscripts and superscripts and followed by individual variable letters which are small letters that we can pick from the part of the alphabet beginning with "x." Of course, the following phrases are not well-formed in PL. Some of them are predicate letters, or individual variable letters, but none is a well-formed formula of PL.

F, H^2, G, xx, xy, R^2z, a, dc, z

Our stipulated grammatical rules do not permit self-standing predicate symbols, self-standing individual variable symbols, or self-standing individual constant symbols. Let us also consider the reason why an expression like "R^2z" is not well-formed. Given the binary character of the predicate letter, as indicated by its superscript, two individual variable or constants symbols (or two mixed variable-constant symbols) ought to follow. This is true even if the superscript is omitted while it is given that the predicate letter stand for a binary predicate.

Now we proceed to construct the grammar for the symbols for the quantificational expressions (also called quantifier symbols.) Let us start with the following well-formed expressions:

F_1^1x, $F_1^1x_1$, F_1^2xy, $F_1^2x_1x_2$

The quantifier symbol - be it the universal or the existential quantifier symbol - must be added to the left of this expression in accordance with the grammatical rules of PL. Indeed, we will recognize the following expressions as well-formed in accordance with the grammar we introduced above:

$\forall xF_1^1x$, F_1^1x, $\forall x_1F_1^1x_1$, $\forall x\forall yF_1^2xy$, $\forall x_1\forall x_2F_1^2x_1x_2$

The same conventions apply for formations that have existential quantifier symbols or mixed universal and existential quantifier symbols. Thus, the following are all well-formed formulas of PL.

$\forall z\exists yFyz$, $\forall x_1\forall x_2Rx_1x_2$, $\exists xFx \vee \sim \forall yGy$, $\sim (\exists xFx \cdot \sim \exists yFy)$,
$\forall x(Px \vee \sim Px)$

Suppose we start with quantifier-free formulas, like the ones on the first row above, and then add the quantifier-variable symbols. E.g.

$\forall x$, $\forall y$, ..., $\exists x$, ..., $\forall x\exists y$, ..., $\exists x\forall y$, ...

Formation rules require that the quantifier is attached only when there is in the formula an individual variable like the one attached to the quantifier symbol. Thus, the following are not well-formed formulas:

$\forall zp$, $\forall xGy$, $\exists x\exists yRxx$, $\exists z(Fx \supset \sim Fx)$, $\forall x\forall yFx$

But we can have individual variable symbols that are not governed - so to speak - by any quantifier. For instance, the following are well-formed formulas:

Gxx, Fxz, Pw, $\exists xRxz$, $\forall x(Fz \cdot \sim Gx)$

When we have an individual variable that is governed by a quantifier with this variable next to it, we say that the variable is <u>within the scope</u> of that quantifier. This is an important concept and we need to become familiar with how to identify immediately what variables fall within or under the scope of which quantifiers. A few examples are given below. Some additional terminology is needed. When an individual variable is within the scope of some quantifier, then we say that the variable is <u>bound</u> by that quantifier; we can also say that the variable is simply bound - which will turn out to be an important element when we consider semantics. An individual variable that is not bound by any quantifier whatsoever is called unbound or - more commonly - <u>free</u>.

Notice that, from the standpoint of mere grammar, formulas can be well-formed even if they have free variables.
In ⌜ $\forall xF^2xa$ ⌝ the only individual variable is bound by the universal quantifier ∀; the individual constant is not the kind of symbol that can be within the scope of a quantifier.
In ⌜ $\exists x\forall yR^2yx$ ⌝ y is within the scope of - or bound by - ∀ and x is within the scope - or bound by - ∃.
In ⌜ $\forall zF^2zz$ ⌝ the second occurrence of z is free; it is not within the scope of any quantifier.
In ⌜ $\exists w\exists u(Fw \cdot \sim Fu)$ ⌝ w is bound by the first occurrence of ∃ and u is bound by the second occurrence of ∃.
In ⌜ $Fy \supset \forall z(Gz \equiv Hyx)$ ⌝
both occurrences of y are free while z is bound by ∀.

- *We disallow <u>vacuous quantifier symbols</u>.* A quantifier symbol is vacuous when it binds no occurrence of its variable. It is crucial to remember that we can have unbound variables in formulas that are well-formed; what the present restriction forbids is having quantifier symbols that do not bind their indicated variable - and this means that they do not bind any other variable either. The following are examples of vacuous quantification and, as such, register as nonsense within our formal system. Other systems of predicate logic may, and do, allow for vacuous quantification. Examples of vacuous quantification are given below:

$\exists xFz$
$\forall x(p \supset \sim p)$
$\forall x\forall yFx$

If we want to indicate, in our metalanguage, that we have quantification on a well-formed formula of our system, we may write:
⌜ $\forall x\varphi$ ⌝, ⌜ $\exists x\varphi$ ⌝, etc.
Having disallowed vacuous quantification, the above kind of expression has a special restriction affecting it. We mean now that the variable of the quantifier (x in the examples above) occurs free in φ.

This means that x is free in φ as it is, before the addition of the quantifier; following addition of the quantifier, the variable then becomes bound.

- *We disallow clashing quantifier symbols*. Clashing, in the relevant sense, occurs in the following instances:

a. When a quantifier variable does not agree with the variable it is supposed to bind. This constraint on the grammar means that we immediately relegate expressions like the following to the bin of ill-formed or grammatically incorrect (nonsensical) expressions. Rather than consider these to be well-formed but failed attempts to bind variables, we simplify things by disallowing this type of clashing.

WRONG: ∀xFy, ∃zFx, ∀x∃yRyy, ∀x∀y(Fx ⊃ Gw), ...

b. When the same individual variable symbol appears as accompanying more than one universal or existential quantifier symbol in the same simple component. (A simple component is one that has no logical connective in it.) We do not prohibit occurrence of the same variable across simple components of the formula.

WRONG: ∀y∀y∃wF^3yyw, ∀x∃xGxy, ∀x∃xRxy ≡ ∃y∀yRyy, ...

The following are grammatically correct because we allow for the same individual variable letter to occur across different simple components of a formula.

CORRECT: ∀x(Fx ⊃ Gx) ⊃ (∀xFx ⊃ ∀xGx), ∀x∃yRxy ∨ ∀x(Fx ∨ Gx), ∀x∃yRxy ≡ ∃y∀xRxy, ...

Although this is permitted, it is not mandated. Hence, it is also grammatically permissible to vary the universal quantifier's variable symbol across components too. Thus, the following are correct too.

CORRECT: ∀x(Fx ⊃ Gx) ⊃ (∀yFy ⊃ ∀zGz),
∀x∃yRxy ∨ ∀w(Fw ∨ Gw), ...

c. For the existential quantifier we make the additional constraining provision that its accompanying variable letters accompanying monadic predicate letters can never be repeated, not even across different simple components of the formula. We could actually skip this restriction but it simplifies things for what happens when we get to substitutions of individual constants into quantified formulas.

WRONG: ∃xFx · ∃xGx, ∃wFw ⊃ ∃wFw, ...

CORRECT: $\exists x_1 F x_1 \cdot \exists x_2 G x_2$, $\exists y F y \supset \exists w F w$, ...

We can consider constraints like the above as amendments to the grammar of PL. You may have noticed that we seem to be stealing glimpses of what lies ahead - the semantic accommodations for our predicate logic system. A characteristic of PRED which we do not find in a language like PROP (a Propositional Logic idiom) is that amendments to the rules of grammar may be needed for certain reasons. This should not be in any way disturbing since a grammar is, anyway, a formal restriction on how to manipulate the symbols of the formal language. Of course, the amendments should not be inconsistent with the other rules or with any consequences following from applying any of the other rules of the grammar.

I.2 PRENEX FORMS

For certain applications of First-Order Logic, and for the extraction of metalogical results, a conversion is required, which takes well-formed formulas of a formal language like PRED and transforms them systematically into logically equivalent forms that are known as being in Prenex Form. It can be shown that every well-formed formula of PRED has at least one equivalent formula that is in prenex form. In a formula that is in prenex form, all the quantifiers have been extracted from within parentheses and placed, the one after the other, in the front of the formula and in order – from left to right – that is determined by the relative length of the quantifier symbols' scopes: thje largest-scope quantifier symbol is the first one in the left, and so one in diminishing order of relative scope length. The order in which the quantifiers are placed in front parallels the order in which their bound variables appear in the so-called matrix – the formula that remains if all quantifier symbols are removed, assuming that all the variables were bound, so that, upon removal, all variables would then be free. Thus, where "Qx" is generally used as a metalinguistic symbol for either a universal or existential quantifier symbol, a prenex form can be represented as follows:

$$Qx_1Qx_2...Qx_n\ (\_\_\_\_\_\_\_\_(---)\_\_\_\_\_)$$

The task requires instructions for extraction of the quantifiers so that they all come in front and become iterated in the same order in which their dependent variables appear from left to right. The task of extraction is not straightforward; it is not to be assumed that a quantifier comes out in front exactly as it is; conversion to the other available type of quantifier (considering that we only have two types of quantifier symbols) may be needed. We show now, briefly, the equivalences on which the extractions rules are based, followed by a few examples. An important directive is that the variables accompanying the quantifier symbols need to be distinguished when the symbols are placed in front. Our grammatical formation rules allow use of the same individual variable symbol accompanying quantifier symbols but, when the front placement is enacted, the variable symbols need to be disaggregated. To this effect, we enforce a different variable symbol for each quantifier and, of course, as we implement this maneuver, we need to make sure that the variables bound by the corresponding quantifier symbols are also changed to match the new quantifier-symbol variables.

$$\_\_(\forall x\varphi \supset ---) \equiv \_\_\exists x(\varphi \supset ---)$$
$$\_\_(\exists x\varphi \supset ---) \equiv \_\_\forall x(\varphi \supset ---)$$
$$\_\_(--- \supset \forall x\varphi) \equiv \_\_\forall x(--- \supset \varphi)$$
$$\_\_(--- \supset \exists x\varphi) \equiv \_\_\exists x(--- \supset \varphi)$$

We can approach this as a matter of first converting implications to inclusive disjunctions and then observing the following distributive equivalences regarding quantifier symbols distributing over disjunctions and conjunctions. Subsequently, we can convert back to implicational formulas. We should find that we obtain the results above.

$$\_\_(\exists x\psi \vee \exists x\varphi) \equiv \_\_\exists x(\psi \vee \varphi)$$
$$\_\_(\forall x\varphi \cdot \forall x\psi) \equiv \_\_\forall x(\varphi \cdot \psi)$$

For example:

$$(\forall xFx \supset \exists xGx) \equiv (\sim \forall xFx \vee \exists xGx) \equiv (\exists x \sim Fx \vee \exists xGx)$$
$$\equiv \exists x\exists y(\sim Fx \vee Gy) \equiv \exists x\exists y(Fx \supset Gy)$$

We could reach the same result following straightforwardly the rules for the implication symbol given above.

$$(\forall xFx \supset \exists xGx) \equiv \exists x\exists y(Fx \supset Gy)$$

I.3 EXAMPLES

1. The following symbolic sequences are grammatically incorrect or ill-formed in PL.

$\forall x\exists xFxx$
$\exists y\exists y(Py \cdot \sim Py)$
$Fb \equiv \sim \forall w\forall wRww$
$\exists y\forall x\exists z\forall xQyxzx$
$\forall x\exists yf(x)$
$\forall x\ \forall y(= x\ y)$
$f(a, b) \supset \exists xFx$

The following are well-formed. Even if the same individual variable is repeated this does not happen within the same simple component of the formula.

$\forall y(Fy \supset Gy) \supset (\forall yFy \supset \forall yGy)$
$\exists xFx$
$Fb \equiv \sim \forall wRww$
$\exists x\exists y(x = y)$
$\forall x\forall y(f(x) = f(y))$
$F(g(a))$

2. The following sequences of symbolic expressions are grammatically correct by the rules of our predicate logic language PL. Another way of saying this is that they are well-formed formulas or they are members of the set of WFFs. A symbolic expression can be well-formed even if it does not express a proposition – as the case is when it has individual

variables in it, which are not bound by any quantifier. Thus, grammatical formation and propositional symbolization go in separate ways: it is possible to have a propositional function or open sentence - a formula that is well-formed but does not express a proposition.

$\exists y A_1^1 y$

$(Gyy \vee Hzz) \supset \exists x \exists y(Rxy \cdot Ryx)$

We don't have clashing quantifiers as it is; if quantifiers were introduced, however, to form another formula, caution would be required. The following, obtained from the preceding by binding free variables, is not a well-formed formula:

$\forall y \exists z(Gyy \vee Hzz) \supset \exists x \exists y(Rxy \cdot Ryx)$

But the following is a well-formed formula. Notice how we re-lettered individual variables to avoid quantifier clashing.

$\forall y \exists z(Gyy \vee Hzz) \supset \exists w \exists v(Rwv \cdot Rvw)$

3. The following symbolic expressions are not well-formed. It is explained parenthetically what the reason is for excluding the formulas from the set of WFFs (well-formed formulas) of PL. (We can still call them "formulas," as we also call them "expressions" or "symbolic phrases," etc., but they are *not well-formed formulas* or wffs.)

4. Examples of conversion of well-formed formulas to equivalent well-formed formulas in prenex form. Explicating the steps: (a) we draw on the definition of logical equivalence as conjunction of an implication and its converse - which we know from Propositional Logic. (b) next, we apply the rules for extracting quantifier symbols from parentheses in a few steps.

$\forall x(Fx \equiv \exists y Rxy) \equiv$

$\forall x((Fx \supset \exists y Rxy) \cdot (\exists y Rxy \supset Fx)) \equiv$

$\forall x(\exists y(Fx \supset Rxy) \cdot (\exists y Rxy \supset Fx)) \equiv$

$\forall x(\exists y(Fx \supset Rxy) \cdot \forall y(Rxy \supset Fx)) \equiv$

$\forall x \exists y((Fx \supset Rxy) \cdot \forall y(Rxy \supset Fx)) \equiv$

$\forall x \exists y \forall z((Fx \supset Rxy) \cdot (Rxz \supset Fx))$

$\forall y(\exists xRxy \supset \exists zRzy) \equiv \forall y\forall x(Rxy \supset \exists zRzy) \equiv$
$\forall y\forall x\exists z(Rxy \supset Rzy)$

$\forall x\forall yRxy \supset \forall z(\exists wRzw \vee Rzz) \equiv$
$\exists x\exists y(Rxy \supset \forall z(\exists wRzw \vee Rzz)) \equiv$
$\exists x\exists y(Rxy \supset \forall z\exists w(Rzw \vee Rzz)) \equiv$
$\exists x\exists y\forall z\exists w(Rxy \supset (Rzw \vee Rzz))$

5. We convert given well-formed formulas (wffs) of PRED into logically equivalent wffs that are in prenex form. We show conversions, in some cases, step by step, with a view to stimulating interest in committing to memory conversion rules of the standard predicate logic.

a. $\forall x(\exists yRxy \supset \forall z(Rxyz \equiv \sim Rxz)) =>$
$\forall x\forall y\forall z(Rxy \supset (Rxyz \equiv \sim Rxz))$

b. $\sim (\exists xFx \supset \exists yGy) => \exists xFx \cdot \sim \exists yGy => \exists xFx \cdot \forall y \sim Gy =>$
$\exists x\forall y(Fx \cdot \sim Gy)$

c. $\forall x((\exists y(x = y) \cdot \exists zRxz) \supset \forall wRxw) =>$
$\forall x(\exists y\exists z(x = y \cdot Rxz) \supset \forall wRxw) =>$
$\forall x\forall y\forall z\forall w((x = y \cdot Rxz) \supset Rxw)$

d. $\exists x \sim \exists yRxy \vee \forall x\forall wRxw => \sim \exists x \sim \exists yRxy \vee \forall z\forall wRzw =>$
$\exists x \sim \exists yRxy \supset \forall z\forall wRzw => \exists x\forall y \sim Rxy \supset \forall z\forall wRzw =>$
$\forall x\exists y\forall z\forall w(\sim Rxy \supset Rzw) => \forall x\exists y\forall z\forall w(Rxy \vee Rzw)$

e. $\exists x(\exists yRxf(y) \equiv \forall wRf(w)x) =>$
$\exists x((\exists yRxf(y) \supset \forall wRf(w)x) \cdot (\forall wRf(w)x \supset \exists yRxf(y))) =>$
$\exists x(\forall y\forall w(Rxf(y) \supset Rf(w)x) \cdot \exists w\exists y(Rf(w)x \supset Rxf(y))) =>$
$\exists x\forall y\forall w\exists w\exists y((Rxf(y) \supset Rf(w)x) \cdot (Rf(w)x \supset Rxf(y))) =>$
$\exists x\forall y\forall w\exists w\exists y((Rxf(y) \equiv Rf(w)x))$

I.4 EXERCISES

(1) Identify the free and the bound variables of the following well-formed formulas of PRED and also identify the quantifier within whose scope each bound variable lies.

$\forall xGxy \equiv \sim \exists y(Hy \equiv \sim Hy)$
$\exists y\exists z\exists w(B^3yzw \equiv \sim (R^2zy \cdot R^2wy))$
$\forall x_1\forall x_2Rxx_1 \cdot \forall x_2\forall x_1Rx_2x$
$\sim (Fxy \equiv Fyx)$

(2) Which of the following sequences of symbols are well-formed formulas of PL?

F^2ab
Faa
$Pax \vee Rxb$
$\forall x\forall y\exists z((Fx \cdot Fy) \supset \sim Fz)$
Fb_{22}
$H^3{}_{11}abc$
$Hxyy$
F^3ac
$Fabc \cdot \sim Facd \cdot (\sim Fbb \vee Faa)$ [Clue: can the arity of a predicate symbol be unsettled or variable? Contrast the following two formulas.]
$F_1abc \cdot \sim F_1acd \cdot (\sim F_2bb \vee F_2aa)$
$G^3xyz \supset \sim G_1{}^2ab$
$\sim Gc \veebar (Fa \cdot Fb)$
$Fb \equiv (Fc \vee \sim (Fd \cdot \sim Fa))$
$\exists w_1\exists w_2\forall w((Pw_1 \vee Pw_2) \supset Q^2ww)$
$Px \supset \sim (Px \supset \sim Px) \vee \sim Px$ [Consider the use of parentheses: is it sufficient to remove ambiguity?]
$\prod u(Hu \vee \sim Hu)$
$\exists xGx \rightarrow Gb$
$\sim Fa_1 \equiv \sim \sim (Gb_2 \cdot \sim (H_1c_3 \vee H_2c_4))$
$\forall FFx$
$\sim \forall x \sim Fx$
$Fa \supset \exists xFx$

$\forall x(Fx \supset \sim Fx)$
$Fx \equiv \sim Ga$
$\exists xFx \supset \sim p$ [Clue: we have allowed retention of the full symbolic resources and incorporation of the grammar; some texts don't allow that but we do.]

(3) Convert the following well-formed formulas (wffs) of PRED to logically equivalent wffs that are in prenex form. Some might be in prenex form as given.

$\forall x\forall y(\forall zRxyz \supset \exists zRzyx)$
$\exists x(\exists yFy \supset Fx)$
$\forall x(\exists yRxy \cdot (\exists zFz \supset \exists yRyy))$
$\exists x\forall yRxy \supset \forall x\exists yRxy$
$\exists xLxx \equiv \forall y\exists x(x \neq y \cdot Lyx)$
$\sim \exists xFx \vee \forall y(Fy \supset Ray)$
$\exists w(Fw \supset \forall z(Rzw \supset Rwz))$
$\forall x(Ff(x) \supset (f(x) = a))$
$\exists x(Rxt \cdot \forall z(Rzx \supset z = t))$

I.5 SEMANTICS FOR PREDICATE LOGIC PL

Now we will pursue a way that allows us to speak of logical meaning - i.e. reference - so that we can tell tales about the objects in our formal language; this, then, allows us to evaluate formulas not only as grammatically well-formed (for which we retain the grammatical rules specified above), but we will also be able to determine the truth values of propositional formulas as to whether they are true or false. This semantic analysis is based on an approach that constructs models; each model is, strictly speaking, an abstract item that comprises a universe or domain of the discourse and an interpretation – also called signature or valuation. How is this to be done? Before we turn to the detailed study of models for PROP, we must take note of an important development we on which we are already able to comment.

In our study of the formal language PROP for Propositional Logic, we could agree to call well-formed formulas propositional formulas or formulas indiscriminately as the two coincided; but in Predicate Logic, there is a distinction to be drawn and it is crucial to understand the underlying motivation for drawing this distinction. In a formal language like PROP, for Predicate Logic, it is possible to have a grammatically correct or well-formed formula which does not express a proposition and, therefore, is *not* a propositional formula. Grammar and semantics part ways – but it is also to be understood that the semantic modeling also relies on syntactical or grammatical specifications for its formal transactions. Consider the following expression which registers within PRED - and it is, in fact, grammatically correct:

Fx

Our syntactic (grammatical-formation) choice was to use individual variables attaching to predicate letters. Semantically, such variables are like the pronouns of ordinary language but with the important difference that they are deliberately construed as ambiguous. While "she is a student" may or may not have an ambiguous pronoun in "she" depends on the context of utterance but "x is a student" in our grammatical construction has "x" as systematically ambiguous. We did this so that we can set up the formal construction of quantified phrases: "x is a student" becomes "at least one x is such that x is a student" if the binding of the individual variable is to be done by the existential quantifier. From a strictly grammatical point of view, the rules for attaching quantifiers to other expressions, as directed, are blind to motivations based on logical meaning. In symbols, the grammatical transformation can be represented as follows:

Fx => ∃xFx

Speaking semantically, the phrase "at least one x is such that x is a student" is a proposition. It is the kind of entity that has a truth value. "x is a student", however, cannot have a truth value given the fact that the individual variable does not have a logical meaning. This means that the individual variable, being systematically ambiguous, has no referent or object it denotes (given that logical meaning is a matter of reference in extensional logic.) Not being able to express a meaning, when the semantic interpretation comes in, the expression ⌜ Fx ⌝ does not make a proposition. We call it an open sentence or propositional function or sentential function. We notice closely that in ⌜ Fx ⌝, x is a free or unbound individual variable but in ⌜ ∃xFx ⌝ the individual variable is not free but bound. Generally, *if any individual variables are free in a formula, its semantic rendering is regarded as an open sentence and it does not have a truth value.* Only if all variables in the formula are bound - or only if there are no free variables - we have a proposition or propositional formula. These considerations account for the fact that logic textbooks attach emphasis on the distinction between free and bound variables and drill students to enable them to identify which variables are free and which are not.

Once again, we notice that open sentences are actually grammatically well-formed. Hence, we have a distinction between well-formed formulas and propositional formulas in predicate logic. We did not have this distinction in Propositional Logic: you can confirm that any well-formed formula of PROP - our Propositional Logic idiom - is also a propositional formula.

We have access only to two quantifiers in standard predicate logic. The universal quantifier, symbolized by "∀", and the existential quantifier symbolized by "∃." The universal quantifier is to be understood semantically as follows:

> ∀xFx: all items in the given domain are F (or have the F-property, or F.)

This means that for every item in the given domain, that item is in the extension specified in the model for F. The signature (or valuation or interpretation) that is comprised by the model will have to specify extensions for all the predicate symbols in the model. Extensions are sets. It is foundationally important that the theoretical machinery of sets is put to the task. A deeper philosophic question about what kinds of things we are talking about within the formal setup of a model will require taking account of the set-theoretic apparatus that is utilized.

The extension of a predicate symbol (also called non-logical constant) is the set of all, and only, the objects of the model's domain which are presumed, in the model's valuation, as having the property represented by the predicate letter. The domain of the model is itself the set of all the objects that are presumed to belong to the model. There is one restriction: that the model's domain cannot be the empty set – although it may well have some size of infinity.

∃xFx: at least one item in the given item is F (or has the F-property, or Fs.)

This means that there is at least one item in the given domain is in the extension specified in the model for F.

In our grammar we stipulated that the quantifier symbol is accompanied by exactly one individual variable symbol from the set {x, y, z, u, w, v, ..., x_1, ...}. We made other grammatical arrangements. Suppose now that we have an open sentence like ⌜ Fx ⌝. This can be called <u>a matrix</u> on which subsequent quantification can be applied. There are exactly two ways for obtaining a propositional formula from this matrix.

a. One way of obtaining a propositional formula is by applying one of the two quantifiers. Thus,

Fx => [∀x] Fx => ∀xFx
Fx => [∃x] Fx => ∃xFx

In applying the quantifier on the matrix, we could not have used any other variable letter besides the one given in the matrix; this is mandated by the way we have constructed our grammar.

b. Another way is by replacing the free individual variable by an individual constant.

The individual constants of predicate logic are semantically understood to be like names: attention is needed is marking that these names are simply labels or tags - they are not supposed ever to have any significant connotations like, for instance, nicknames of ordinary language may have. The names of our PRED system simply label specified objects. Since names, so understood, cannot be ambiguous, an expression like "the object named a has F predicated of it", symbolized as "Fa" cannot be ambiguous and, as such, it carries a truth value and is a proposition. It is inevitable, then, that our semantic setup should assign exactly one object to each name or individual constant. In other words, the assignment of objects to names is functional. Notice the symbolization we use (in our metalanguage) to indicate substitutions into matrices. For instance, substitution of the individual constant a for x is used in the first example to produce a propositional formula.

$Fx \Rightarrow [a_j/x]\, Fx \Rightarrow Fa_j$

The above can be easily generalized to multiple free variables. If one or more free variables appear in the formula, it is considered a matrix. [Of course, given the specifications of our grammar, matrices, as defined, are still well-formed formulas of our system.]
(Remember that the matrices themselves are well-formed formulas of predicate logic but they are not propositional formulas.) At this point, we don't have to specify that substitutions into free variables have to be uniform. Our present concern is just moving from open sentences (matrices) to propositional formulas. But we will practice uniform substitutions (or uniform replacements) as well because this is important for reasons we will discuss.

Examples of binding multiple free variables or replacing free variables by constants in matrices to construct propositional formulas follow.

$Rxy \Rightarrow \forall x \forall y Rxy$

$Rxy \Rightarrow \forall x \exists y Rxy$

$Rxy \Rightarrow \exists x \exists y Rxy$

$Rxy \Rightarrow \exists x \forall y Rxy$

$Rxx \supset Rzw \Rightarrow \forall x \exists z \exists w (Rxx \supset Rzw)$

$Rax \Rightarrow Rab$

$Rax \Rightarrow \exists x Rax$

$Rxa \Rightarrow \forall x Rxa$

$\sim Ray \vee (Rbz \supset Rzw) \Rightarrow \sim Rab \vee (Rbc \supset Raa)$

Of course the different ways of binding result in different meanings. We will be turning to translations soon. The propositions expressed, for instance, by $\ulcorner \forall x \forall y Rxy \urcorner$ and $\ulcorner \forall x \exists y Rxy \urcorner$ do not have to have the same truth value; hence, they have different logical meanings. They also translate different, given a lexicon, to and from English.
tr($\forall x \forall y Rxy$) = any two things in the domain are R-related.
tr($\forall x \exists y Rxy$) = anything in the domain has at least one thing to which it is R-related.

Even though we do obtain propositional formulas regardless of what constants we use to replace which free variables, it is crucial in many procedures that the *same constant* is substituted throughout the matrix for the same individual variable. For example,

Fx => [a/x] => Fa
Rxx => [a/x] => Raa
Rxx => [b/x] => Rbb

Hence, if we are pursuing uniform substitution or uniform replacement of the free variable, we cannot have:

Rxx => Rab (!)
Fx · Rxyz => [a/x, b/y, c/z] => Fa · Rcbc [!]

We can easily appreciate the significance of uniform substitutions. We can look back to our Propositional Logic first. Suppose that we have a logical form, for which, as we know, we use small letters:

$\sim p \vee \sim (q \supset r)$

Now, suppose that we want to assess propositions expressed by English sentences - symbolized per our grammar by capital letters - with respect to what logical forms they instantiate. This is crucial because the characteristics of the instances are considered to be due to the characteristics possessed by the form. We can think of forms as TYPES and specific instances of those forms as TOKENS. As an example, think of a \$1 token pass for a train as an instance of the type "\$1-pass-for-train." Or think how the word "blob" has four token letters but three letter types (b, l, and o.) Now, to assess if a propositional token is an instance of a type or form, we must rely on checking if we can make uniform substitutions into the form to produce exactly the propositional token we have. If this is not possible, then the propositional token does not match with the given form. Consider, again, the given form above and the following propositional tokens and see which ones are and which ones are not tokens of this form. We provide the justification by showing the uniform substitutions for the variables p, q, and r, which yield the token propositions.

$\boxed{\sim p \vee \sim (q \supset r)}$

$\sim A \vee \sim (B \supset C)$ A/p, B/q, C/r

$\sim W_1 \vee \sim W_2$ W_1/p, W_2/q $\supset$ r

$A \vee (B \supset C)$

This, however, is not a token of the form. No uniform substitutions into the form can produce this token.

In the case of predicate logic languages like the one we have constructed here, the uniform substitutions replace individual variables by individual constants so that the same constant is used to replace the same variable. Predicate logic has resources for looking into the analyzed structure of propositions. We can now render analytically the internal structure of the proposition expressed by a sentence like:

(S) Everyone who loves himself or herself loves someone.

We will pursue translations subsequently but here is a brief attempt, noting already in this early stage that universal statements have an implicational (if-then) character to them.

For all x, if x loves x, then there is at least one y such that x loves y.

$\forall x(Lxx \supset \exists yLxy)$

Now, semantically, non-uniform substitutions would alter the sense or meaning of the quantificational phrase. For instance,

$Lab \supset Laa$

means something different and is an instance of "if someone loves someone else, then he or she loves himself/herself." On the other hand, we can have a uniform substitution as:

Laa ⊃ Lab

Now, the meaning has not been altered. The quantificational phrase does not specify whether the object of the person's love can be someone else; the way it is phrased, this actually seems implied.

*** Let us then elaborate on the crucial issue of substituting constants into given quantificational formulas. If we are given quantificational formulas as true or false, we notice, intuitively, that universal quantification gives us license to substitute any individual constant letter. Essentially, as we do this substitution we eliminate the universal quantifier - which we may mark by "∀Elim".

∀xFx => [a/x] => Fa (∀Elim)
∀xFx => [b/x] => Fb (∀Elim)
etc.

Existential quantification does not give us such license but it allows us to substitute *some* individual constant letter since we are told that at least one item in the domain is in the extension of the predicate. This substitution eliminates the existential quantifier. But it becomes imperative to use a *new constant* - one that has not been used for any other substitutions. And we must thereafter track the constant we used for substitution in the step of eliminating the existential quantifier to make sure it is not used in other existential eliminations. Of course, this constant can be used in universal quantifier eliminations; this is because anything may be substituted for universal elimination.

∃yGy => [a/y] => Ga (∃Elim);
∃zFz => [b/z] => Fb (∃Elim)

We must use a different constant in the second elimination. It is not difficult to see the logic behind this. Existential quantification simply informs us that *some* - at least one - item in the domain has the property. We are told nothing about which object and we do not have the license to assume that we know anything about this object. The object must have some name (see below), and, so, there is *some* constant that labels it. That's all we know. When we have to perform a subsequent existential elimination, we have no right whatsoever to assume that the same object of the domain must have the second property.

Objects in the given domain do not have to have only one name tracking them. There can be more than one names labeling or tracking the same object. You must recall that we have an identity symbol in our grammar for PL. You can see how this plays out here. If a_1 and a_2, for instance, are two individual constants or names that both name the same object, then we have that the value of the proposition "$a_1 = a_2$" is true. We must make a deliberate effort to understand what this means. "The object named by a_1 is identical with the object named by a_2." This means that we are talking about one object. It does not mean that we have two objects that are "identical" in some other sense - in usage or purposes, for instance. Logical identity, then, is a matter literally of one object being referenced or denoted by the same name. The names - or individual constants - are then considered identical and this also means that the proposition "name a1 is identical with name a2" is true in the specified context in which this setup is presented. The phrase "the referent of a_1 is the same as the referent of a_2" is a proposition, of course; it does not have any free individual variables, it only has individual constants that are never ambiguous in their meanings (which are their references).

"---is the same as___" is a relation; it is a relation with special characteristics. It is a binary relation. In our semantic setup for predicate logic, binary relations are to be treated as semantic renderings of binary predicates. This simple but revolutionary extension of the concept of monadic predicate to n-ary predicate had eluded giants of antiquity like Aristotle; its incorporation into logical analysis made it possible for formal logical systems to exceed in power anything that had been done before.

Next, we need to give a formal and precise definition of what a model is for our purposes. Well, the model is itself, again, an abstract item - like the kinds of things Mathematics deals with - even though the model will allow us to tell tales, as promised. This is not a problem. We have flexibility in choosing domains of items so that we could be talking about things we can relate to in our daily experience or about objects as abstract as, for instance, the positive integers or geometrical shapes. We will now define what a model is. Let us start by writing out, in our metalanguage, what we mean by "model" before we provide specifics about what is involved.

$$\mathcal{M} = \langle D, \surd \rangle$$

A model is an ordered pair. We can see this above because the symbolism "<---, __>" is used in set theory for signifying an ordered pair. The pair is ordered because the order in which the items are constructed matters. The first item, to the left, symbolized by "D" is called the domain or universe of discourse and must be generated first in model-construction; the definition of what is symbolized by "√" - soon to be elaborated - comes next because the valuation assignments that are collectively symbolized by "√" must be defined *over* the domain D. A restriction we impose is that the domain cannot be empty although it can be infinite.

$$D \neq \emptyset$$

Thus, a model is an ordered pair comprised of a domain and what we call a cluster of valuation functions. The latter is generally called a valuation - in various texts it is also called an interpretation or a signature.

Strictly speaking, a model is the abstract object we have defined with respect only to the propositional formulas which it satisfies: this means that those propositional formulas receive the truth value True in the satisfying model. Since True is the negation of False, the model also satisfies the negations of all propositional formulas that are not true but false (are not satisfied) in the model. Indeed, we can easily prove that every propositional formula must be either true or false in a given model in which it is constructed given the model's signature. If we insist on calling the abstract object a model only vis a vis the formulas it satisfies, then we will have to call the pair <D, √> by some other term – the most common one being interpretation. If this is the case, we should refrain from calling the signature also interpretation. It is unfortunate that we do not find universally uniform nomenclature in logic but the remedy of disambiguating precisely is well within reach.

The domain of discourse is itself a set. It has to be the set of objects about which the model allows us to talk. The valuations, as we will soon see, are defined over those objects. A domain can be restricted or unrestricted. Let us start by explaining what an unrestricted domain is. We will use a different letter to symbolize exclusively the unrestricted domain. Of course, the unrestricted domain too, like any domain, is a set of objects.

$\Omega = \{x \,/\, x \text{ is anything that exists}\}$

The existential stipulation should not be understood in a metaphysical sense. Fascinating puzzles have titillated the philosophic imagination for long eons but the formalism we are using compels us to certain principles about existence – not because this can be taken as a proof about what exists and what doesn't but simply as a consequence of how our modeling approach works. We will have more to say about the commitments we incur about existence when we use this formalism.

The unrestricted domain, for instance, is not a heuristic tool – it is not the result or an accompaniment to empirical discoveries about what things or kinds of things exist. We may think of the choice of unrestricted domain for modeling as telling us that "anything that can be talked about is in the domain" or, since this is rather vague, "the choice of domain does not commit you to any specific kind of entity." This means that we will have to use predicate symbols, interpreted linguistically as "person" or "thing" or "dog" or whatever, to add concrete specificity to our attributions of predicate properties. If, on the other hand, we have a restricted domain, which is restricted, for instance, to persons, we don't need a predicate letter to be interpreted as "person."

The unrestricted domain has anything in it. As we have seen, if this is selected for the model, there are consequences about how to translate from English into our formal language. In that case, all characteristics (like "being a person" or "being an object of some sort" need to be in the model as logical predicates.

For instance, here is how we would translate over an unrestricted domain the English sentence (to be precise, the proposition expressed by the English sentence),

(S) Everyone loves at least someone.

Analysis:

For all x, if x is a person then there is at least one y such that y is a person and x loves y.

In our established symbolic notation:

$\forall x \exists y(Px \supset (Py \cdot Lxy))$

or, the logically equivalent propositional formula,

$\forall x \exists y((Px \cdot Py) \supset Lxy)$

But in a domain that is restricted to only persons,

D = {x/ x is a person}, our translation is:

$\forall x \exists y Lxy$

The domain for a model is always specified. We can think of it as being a matter of context-sensitive specification, which, happily, seems to be the case also for everyday usages of a language like English.

Examples of restricted domains are given below. A restricted domain has to be a proper subset of the unrestricted domain. There must be some things that are not in the restricted domain although they are considered as members of the absolute or unrestricted domain. For restricted domains we retain the generic domain-symbol "D" while we use "Ω" for the unrestricted domain.

D1 = {x / x is any dog alive at any time during the period 1980-2015}

D2 = {1, 2, 3, ...}

D3 = {x / x is a student in a specified class}
D4 = {x / x is any person}
D5 = {♚, ♛, ♜, ♝, ♞, ♟}

Needless to say, the unrestricted domain is infinite. We talk about numbers and they too must be presumed within the unrestricted domain. Regardless as to whether an infinite number of things exists at all times in the universe, the unrestricted domain also contains numbers and those form infinite series. But a restricted domain too can be infinite: this is the case with D2 above, which is infinite but is still a restricted domain since it does not have absolutely everything that is considered existent in it. (Notice that context of existence does not matter: if something existed in the past or has not yet existed does not matter insofar as we allow ourselves to treat this object as an existent for the purpose of telling stories about it.) We cannot handle assorted restrictions – for instance, specifying restriction to dogs and cats in a systematic way and so that we do not need to use predicate letters for the dogs and the cats: that would require sortition of the quantifiers, which means that we would need to have two pairs of universal and existential quantifiers, one for one type of entity and the other for the other type of entity in our sorted domain. This can be done but it takes us beyond the pale of standard logic into an alternative or non-standard predicate formalism.

Notice also that the domain, even though it is an abstract thing (a set), has as elements things or objects. Of course we place in the domain, as in D5, stand-in symbols for the things: we cannot possibly put the things themselves. Now, these are not the names of those things. In fact, there could be as many different ways as you like for naming those things. The important issue here is to label or track those objects: this is why we name them. The names are not to have any deeper significance but they are merely labeling or tagging devices. This role is performed by our individual constants. What a name is and does is yet another intriguing philosophic problem but our formalism treats names as extensionalizing devices: they label or tag objects so we can talk about them, and that is all they do. We must make certain stipulations:

- Every object in the domain is considered as having a name - an individual constant referring to it. No nameless entities are allowed. The labeling is presumed systematic and complete relative to the objects of the domain. A name must have exactly one object it refers to; no ambiguity is allowed to creep in about reference. On the other hand, however, two names can refer to one and the same object. An initial notion that there are two separate objects must then be dispelled when it is stated that "they" are named by one and the same name. We need the symbol of identity in the formal system to indicate this. Thus, co-reference means that the presumed items referred to are identical with each other – there is one, not two items. Our symbolic resources in Predicate Logic do not permit – interestingly enough – stating the famous Leibniz Laws of Identity of Indiscernibles and Indiscernibility of Identicals. For that we need Second-Order Logic resources, which would allow us to symbolize properties or properties (second-order predicate symbols.) The Leibniz Laws determine identity of two initially presumed distinct objects on the basis of sharing all their properties – and also declare two indiscernible objects to be necessarily sharing all their properties. Thus, we would have to be able to quantify over predicate symbols – and Predicate Logic does not make such symbolic resources available; nor do we have the semantic adjustments we need to such an effect.
- Every individual constant in our PRED system must have an object in the domain to which it refers. There are no non-referring constants or names. That would be tantamount to having names that are meaningless (since they have no references) and yet they are counted within the list of labels we have in the model. The unique referent, and only that referent, is the meaning of the individual constant. This we can call the referent of the individual constant and it is the value

given to it by the model's signature or valuation. Thus, the value of an individual constant ⌜ a_j ⌝ is an object in the domain, or belongs to the domain. In set-theoretic language "∈" symbolizes the attribute of belonging to a set. Thus, we have:

$\surd(a_j) \in D$

This is one of the operations that fall under the signature or valuation, which is a cluster of functions symbolized by "√". This valuation-function confers logical meaning to all the individual constants by assigning to them an object of the domain to which each refers. This is a function since the output – the assigned object – is unique for every specified input (a constant symbol).

- The fact that, as we pointed out above, two or more individual constants can refer, or co-refer, to the same object of the domain does not affect the fact that the assignment of objects to constants - the valuation - is a function. Each constant has a unique object it names - even if more than one constants have the same object to refer to.
- Function symbols denote objects in the domain. They can take as inputs individual constants or other function symbols. Thus, function symbols too can co-refer with individual constants or with other functional symbols.

The predicate logic we are studying has a built-in bias in favor of taking its domain objects as existing. We give full disclosure of this without entering into some fascinating philosophic - and logical - issues that revolve around this subject. This has consequences that will become apparent in due time. There is even a philosophic view that the only legitimate and correct way of presenting metaphysics (views about what things and kinds of things may exist) is by showing how to construct a formal language - an idiom of predicate logic - in which the quantifiers quantify over the elements of the given domain. We could choose – as an extra-logical commitment – to defer to the pronouncements of science, for instance, and restrict our domain permissions only to such things as are allowed by science. Then, all such things would be presumed existing – notice, not as the result of some overarching metaphysical inquiry but as a matter simply of deferring to the authority of science. Any other way of plunging into metaphysical speculations about what exists is bound - on this view - to result in dangerous confusion. If this view is accepted, then issues of existence are to be settled simply by context-specific enumerations of the members of the domain of the relevant formal language. For instance, forget some abstruse debate over whether some subatomic particles like mesons exist or not. The fragment of English in which Quantum Theory is conducted is formalizable by a predicate formal system and the domain of that system has mesons in it - and that's all there is to metaphysics. (It is to be doubted that standard logic is appropriate for the formalization but we cannot enter into this matter here.) Notice also that, on this view, predicate logic is considered adequate for translating formally theoretical claims in Mathematics and in the sciences. We will be returning again to this subject. It should be mentioned that linguistically meaningful tropes of speech like "Pegasus does not exist" cannot be translated into our formal predicate language; they are nonsensical because, as we will have a chance to see, every domain object "exists". The notion of domain-relativized existence can be captured by the statement, "for all x, there is a y, such that x is identical with y." This does not sound like a logical or analytic truth but our formalism commits us to it. One might chafe at this consequence of adopting this formalism in the first place. We cannot say that something exists, within the formalism, without stating a logically trivial truth (but this does not seem to be the case in language, with existential statement being obviously contingent); nor can we use the formalism to state that something does not exist on pain of using a nonsensical sentence. We may even take issue with what seems to be a formally specific redefinition of "existence" imposed by the formalism. There are alternative logics but, for our present purposes, we simply lay all cards on the table, so to speak, and proceed with our systematic task. Based on what we have stated, as we move on to account for the treatment of predicates, we may also point out that an existential predicate would be ill-advised insofar as this formal system is concerned. This appears to vindicate a famous principle promulgated by Immanuel Kant, in his rejection of the Ontological Argument for the existence of God, according to which existence is not a logical predicate. The subject is more complicated but we cannot pursue it further here. It has been seminal in the historic developments that led to modern logic that predicate logic seems adequate for the formalization of the language of Mathematics – but, oddly enough, a mathematical principle like that of Mathematical Induction cannot be stated properly without resources of higher-order logics!

Next we must provide for ways of assigning logical meanings to the predicates within a model. The valuation function assigns logical meanings to all the predicates for which we have symbols or letters. The predicates also receive logical meanings or valuations within the model. This is how we do this.

The valuation (or interpretation or logical meaning) of a unary predicate symbolized, generally, by "$\surd(F^1{}_j)$", also called the extension of⌜ F^1 ⌝, is the set of elements of the domain that are considered, in the model, as having the attribute interpreted by the logical predicate. Note that this set is a subset of D; so, it is itself a set of objects of D. It does not have to be a proper subset. It can be the domain itself: obviously, this would mean that all members of the domain have this predicate or attribute. An extension can also be the empty set. In other words, it is not disallowed that we set up a model in which none of the domain's objects have a certain attribute for which we do have a symbol in the model's signature. It is significant that empty extensions are allowed; yet, as we have noted, non-referring constants are not allowed and also domain objects without constants naming them are not allowed either.

This is the case for unary predicates. Thus we have - symbolizing the relation of being-a-subset by "⊆":

$$\surd(F^1{}_j) \subseteq D$$

The powerful insight that applies the mathematical concept of function to the treatment of logical predicates extends to the case of generally n-ary or n-place predicate symbols. This means that we can now manage the logical treatment of relation, a subject that sadly eluded Aristotle and other ancient students of logic. An n-place function, formed syntactically from attaching to the predicate symbol the same number, n, of individual variable symbols, has an extension that is a subset of the nth Cartesian product of the domain by itself. We show examples below:

$\surd(F^2{}_i) \subseteq D \times D = D^2$

...

$\surd(F^n{}_i) \subseteq D \times D \times \ldots \times D = D^n$

We can also speak of anti-extensions of predicate symbols - although this is rarer in the literature. The anti-extension of a predication can be symbolized by "$\surd'(F)$" and is the set of all the elements and only the elements of the domain, which are not in the extension of ⌜ F ⌝. The anti-extension is the set that, in set-theoretic terms, is called the complement of the extension-set for the predicate. The union of extension and anti-extension of the same predicate is the set that has all the elements in the two sets (as we say, it has every element of D that either belongs to $\surd(F)$ or to $\surd'(F)$.) This union should be equal to the domain set - should have exactly all the same elements with the domain set. The overlapping or intersection of $\surd(F)$ and $\surd'(F)$ is the set that has all and only the elements of D that belong in common to both the sets: the intersection set should be the empty set. In other words, the extension and anti-extension of a given predicate letter are mutually disjoint sets: they have no members in common and when their members are all put in a set we get the domain of the model. If it were otherwise, we could not say that "for every x, x is either F or is non-F"; and we could not say that "for every x, it is not true that x is both F and non-F."

The condition that the extension and anti-extension of every predicate letter are mutually disjoint guarantees that we are dealing with standard predicate logic for which the logical principles of excluded middle and non-contradiction are valid.

(Excluded Middle) $\forall x(Fx \vee \sim Fx)$

(Non-Contradiction) $\sim \exists x(Fx \cdot \sim Fx) \equiv \forall x \sim (Fx \cdot \sim Fx)$

To inspect an example, we can have a model as follows. We use subscript "$\mathcal{M}$" to indicate that the valuation is specific to our given model. We could construct alternative models of course, each with different valuations even for one and the same domain and stock of constants and predicate letters. Obviously, two models that agree on both domain and valuation functions are actually one and the same model.
Here is an example.

$\mathcal{M} = \langle D = \{1, 2\}, \surd \rangle$
$\surd_{\mathcal{M}}(a_1) = 1$
$\surd_{\mathcal{M}}(a_2) = 2$
$\surd_{\mathcal{M}}(F_1) = \{1, 2\}$
$\surd_{\mathcal{M}}(F_2) = \emptyset$
$\surd_{\mathcal{M}}(h(1, 1)) = 2$
$\surd_{\mathcal{M}}(f(1)) = 2$

This tells us the following story.
The object metalinguistically denoted by "1" has its name symbolized by ⌜ a_1 ⌝. (Inevitably, we cannot place the object itself anywhere; we must use stand-ins for the domain-objects in the metalanguage.) The object metalinguistically denoted by "2" has its name in the model ⌜ a_2 ⌝. The naming letters are called individual constants in predicate logic systems and we consider ourselves to have an infinite supply or stock of them as needed (with the infinity being that of the positive integers or, as it is called, denumerable.)
The symbol "s" stands for the function of summation, defined over the pair <1, 1>. The denotatum is a member of the domain. The function symbol "f" stands for the successor and denotes, for the number 1, the

number 2. It so happens that the unary and binary function symbols of the model co-refer or have the same denotatum.

The defined predicates in the model are denoted by "F_1" and "F_2" and have logical meanings or valuations (also called extensions) as shown above. We can now reflect on the decisive choice to express properties by means of sets which we call extensions or valuations. Thus the properties that are given interpretations or valuations in the model are presented by means of the sets of the members of the model domain, which have the properties in question. An extension is a valuation function that takes inputs from the predicate symbols set and yields outputs from the set of subsets of the domain (which is called the Powerset of the Domain.)

Now, we can figure out truth values of certain propositions for the valuations specified in the given model. We will go over details soon but, intuitively, we can see how we would answer questions about truth values in this model. Interestingly, truth value assignments are now relativized to models. But logical truth is not because no formula is considered as a logical truth or tautology if it is possible to construct any model in which this formula is false.

$\surd_{\mathcal{M}}(F_1a_2) = T$ because
$\surd_{\mathcal{M}}(a_2) = 2 \in \surd_{\mathcal{M}}(F_1) = \{1, 2\}$

This is because the object referred to by "a_2" indeed belongs to the extension-set or value of the predicate denoted by "F_1". We can write this in the metalanguage:

$\surd_{\mathcal{M}}(F_2a_1 \vee F_2a_2) = F$ because
$\surd_{\mathcal{M}}(a_1) = 1 \notin \surd_{\mathcal{M}}(F_2) = \emptyset$ and
$\surd_{\mathcal{M}}(a_2) = 2 \notin \surd_{\mathcal{M}}(F_2) = \emptyset$

Neither object belongs to the extension of F_2, which in this model is the empty set; this means that both propositions denoted respectively by "F_2a_1" and "F_2a_2" are false; as we know from Propositional Logic, the inclusive disjunction of two false propositions is a false proposition.

$\surd_{\mathcal{M}}(f(1)) = \surd_{\mathcal{M}}(h(1, 1)) = 2$

The value of a function is its referent which is an object in the domain (functions are terms, like constants). The two defined functions of the model have the same referent – hence, they co-refer, which means that identity obtains between them. We keep in mind that function symbols are terms – like individual constants and individual variables. Like constants – and unlike variables – function symbols are assigned valuations by the model's specified signature, with such valuations being in terms of objects of the domain.
Formulas that are evaluated as True in a model are called satisfiable in the model while those evaluated as false are called non-satisfiable in the model. We always have only two truth-values in standard logics; so, notions like satisfiability/non-satisfiability, consistency/inconsistency, and validity/invalidity have a pleasantly simple binary character. Whether this is always the case when it comes to the study of the logics embedded in natural languages is another matter – a controversial one, which falls outside our current scope. A set of formulas all of which are true in a model is called also satisfiable while it takes even one, or more, false formulas in a set to declare it non-satisfiable. In the case of sets of formulas, we can speak interchangeably of satisfiability/non-satisfiability and consistency/inconsistency.

Thus, to recapitulate: a given model's valuation signature $\surd_{\mathcal{M}}$*, defined over domain D, assigns as values (also called interpretations) objects of the domain for constants and function symbols and sets of objects or of pairs of objects for predicate letters.*

Let us examine an example of a relation modeled with the resources of predicate logic.

$\mathcal{M} = \langle D = \{o1, o2\}, \surd \rangle$

There are two objects in the model domain, which are indicated as "o1" and "o2" (obviously, we cannot have the objects themselves in the notation.) We proceed with specifying the model's valuation assignments:

$\surd_{\mathcal{M}}(a) = o1$

$\surd_{\mathcal{M}}(b) = o2$
$\surd_{\mathcal{M}}(R) = \{<o1, o1>, <o1, o2>\}$
The members of the extension have to be ordered pairs. For x being R-related to y, we need a different notation than for y being R-related to x. The order matters. Thus, the ordered pair <o1, o2> indicates that object o1 is R-related to object o2; it says nothing about how o2 may or may not be related to o1. We are approaching this abstractly but texts also like to given concrete examples; there is nothing wrong with that - but there is nothing special either. Considering the binary relation "x-being-the-father-of-y", symbolized as "Fxy," we can reflect that for two names, a and b, the truth of the proposition Fab actually implies that Fba is false.

To be sure, we could arbitrarily construct a model in which they are both true. The abstract way we set up models allows us to do this with impunity. Nothing depends on this either in a sense: relations like that of being-father-of cannot be involved in lists of logical truths. It is a matter of biology, not of logic, that this relation is not symmetrical. Hence, it is not at all destabilizing to our formal intentions that we can in theory build any model we like with the extension of any predicate defined any which way. Nevertheless, we could not claim to have constructed a representative model for the fatherhood property if we were to build a model in which we made this relation symmetrical. Something else to commit to memory already: the only exception is the logical predicate of identity, which we have mandated that we symbolize with "=", and which we could be treating in symbolization as a binary predicate, rendering by "Ixy". For this logical predicate only we will set constraints on what models can be built and cannot be built. For all other logical predicates we may, if we wish, apply what we call meaning postulates that restrict how extensions are built - as we making the fatherhood relation being asymmetric. But with respect to the identity relation we will not leave it up to choice whether to apply meaning postulates or not: we mandate, as is customary, that certain restrictions apply. We will get to all this shortly.

Continuing with our model, we can determine truth values of propositions in the context of this model. We are doing this rather informally and appealing to common sense, something that model-building seems well suited to doing.

$\surd_{\mathcal{M}}(Rab) = T$ –

since <o1, o2> belongs to the extension of R in the model, then the proposition Rab is true. We can think of this as saying that it is indeed true in the model that what is named by a is R-related to what is named by b. Next, try to make sense of the following valuations.

$\surd_{\mathcal{M}}(Raa) = T$; $\surd_{\mathcal{M}}(Rbb) = F$; $\surd_{\mathcal{M}}(Rba) = F$.

What about quantified phrase like the following?

$\surd_{\mathcal{M}}(\forall xRxx) = F$ - this is correct since it is not the case that all objects of the domain are R-related to themselves; one object is so related but the other object isn't. Now account for the following valuations:

$\surd_{\mathcal{M}}(\forall x\exists yRxy) = T$ - reading this as "every object in the domain has at least one object to which it is related in the domain."

$\surd_{\mathcal{M}}(\exists x\forall yRxy) = F$ - this says that "there is at least one object to which all objects are related in the domain", and this is not the case. We will study other approaches to how to analyze such propositions with a view to determining their truth values in given models.

Now let us consider the following:

$\surd_{\mathcal{M}}(\exists xRxy) = ?$

What this well-formed symbolic formula expresses does not have a truth value because it is not a proposition: it is a propositional function or open sentence; this is because it has a free or unbound variable ⌜ y ⌝. Think of this as saying: "there is at least one thing to which y is related in the given model." The ambiguity of "y" is definitive. Unlike in ordinary language, in which usually "he" or "she" or "it" are understood to have a referent in a given context, the individual variables, like y, are systematically ambiguous. They cannot have a referent - hence, they have no value - given the way we have constructed our formalism. Therefore, "there is at least one thing to which y is related in the model" is systematically, and deliberately, ambiguous and its ambiguity is intrinsic to it and it is irremediable. Its truth value cannot be determined ever. It makes sense to say that this is the kind of thing that cannot have a truth value: it is not a proposition. We have called such expressions open sentences or propositional functions. We can treat this expression as a template so that by filling individual constants into it (for the free variable) we then generate propositions. Let us try:

$\surd_{\mathcal{M}}((\exists xRxy)[a/y]) = \surd_{\mathcal{M}}(\exists xRxa) = T$ -

now we have a proposition, since we have no free variables. We can think of this as saying that "there is at least one thing to which the object named by ⌜ a ⌝ is related to in the model." This is correct for this model; hence, the proposition is true in the model. [We could surely build another model in which this proposition would be false. This is exactly what we want to be able to say. The proposition symbolized by ⌜ ∃xRxa ⌝ is not a logical truth or tautology and should not be true in every model. Nor is it a logical contradiction and it should not be false in every model - and it is not.]

Let us now underline some important decisions that have been made with a view to providing the semantics for our predicate logic system. There is bound to be repetition of points made earlier but, this time, we draw attention to certain details. These decisions have important consequences, some of which might not be intuitive. There are alternative ways of constructing formal systems for the standard predicate logic. There are also approaches to the study of predicate logic that result in the construction of systems that are alternatives to the standard predicate logic we examine here.

- Aristotle had assumed that every class has at least one member; or, in other words, that there are no empty classes. This would mean for us that we stipulate that there are no empty extensions in a model. But we have not done this. Following the standard approach in modern predicate logic, we have allowed for possibly empty extensions. The values of the logical predicates (also called non-logical constants) are defined to be their extensions. Hence, we have allowed for logical predicates that have as logical meaning or extension the empty set. To indicate in a model that there are no unicorns, for instance, we must assign the empty set as extension to the logical predicate we can symbolize by "U" and set in the lexicon as "Ux : x is a unicorn." This provision for empty extensions has some counter-intuitive consequences. You should not concern yourself unduly at this point with these matters but you should be forewarned that certain consequences follows. Such results are avoidable or remediable but the correction would involve an alternative approach to the way of constructing standard predicate logic. If we have $\surd(U) = \emptyset$, then any quantificational phrase like

$$\forall x(Ux \supset Fx)$$

for any logical predicate symbolized by "F", will have to be true! Let us examine this.

$$\surd_M(\forall x(Ux \supset Fx)) = F$$

if and only if there is some entity in the domain, named by constant "a", such that $\surd_M(Ua) = T$ and $\surd_m(Fa) = F$.

However, since $\surd_m(U) = \emptyset$, it follows that there is no object whatsoever in the domain for which $\surd_m(Ua) = T$. Hence, for our object, referred by "a", it cannot be that $\surd_m(Ua) = T$ as we assumed in order to make it that $\surd_M(\forall x(Ux \supset Fx)) = F$. This means that we cannot make the generalized implication $\forall x(Ux \supset Fx)$ false! It has to be true in the model in which the extension of the logical predicate unicorn is empty. Another result is that we can never allow ourselves to symbolize an existential proposition that involves co-attribution of two or more predicates as an implication. For instance, let us consider the statement:

Some flying horses are insects.

If we symbolized - for the obvious lexicon - by,

$$\exists x(Fx \supset Ix)$$

then, we have a vacuously true proposition expressed by this formula insofar as the extension of flying horses is empty in a model. (An implication is vacuously true just in case the implication is true because its antecedent is false.) Although we tolerate vacuous implication in cases of universal quantification, as we said above, it seems intolerable that we would have to admit vacuous truth in cases of existential generalization.

Thus, we must always treat existential quantification involving more than one predicates as conjunctions.

$\exists x(Fx \cdot Gx)$, $\exists z(Hz \cdot\cdot Gz)$, $\exists x_1\exists x_2(Fx_1 \cdot Fx_2)$,
$\exists x\exists y\forall z((F_1x \cdot F_{2y}) \supset Gz)$, ...

As already indicated, Aristotle presumed that there can be no empty predicate extensions. Another way of saying this is that Aristotle presumed existential import for all logical predicates. It is arguable that this presupposition of existential import is quite common in everyday linguistic practices. If you are told that "all the sons of John are carpenters", you assume that the extension of the logical predicate "being-a-son-of-John" is <u>not</u> empty. If it were the case that it is empty, the implicative statement "for all persons, if such person is a son of John, then he is a carpenter" would be vacuously true! Recall the definitions of the logical connectives of our Propositional Logic. (The same connectives are retained in setting up the predicate logic.) The implication connective permits vacuous implication: if the antecedent is false, then the implication statement is true.
Regardless of what seems to be the case with everyday linguistic applications, it should be borne in mind that the standard approach to predicate logic, which we present here, does not presume existential import. We may well have empty predicate extensions in our models.

- The domain of a model cannot be empty. It has to have at least one element. In other words, we do not permit ourselves to contemplate the case of an empty universe of discourse. Moreover, we allow ourselves use of exactly one domain: so, to sum up, no empty domain is allowed and no more than one domains are allowed. These restrictions have certain consequences. We cannot, for instance, parcel out our items over two domains - let's say, one for actually existing things and the other for fictitious or alternative-universe entities. So, if we want to speak of the flying horse Pegasus, for instance, then Pegasus is in our domain. "Pegasus," the name, which may be “p" as our grammatically chosen individual constant, refers to or denotes the famous mythological flying horse. We cannot say that "Pegasus does not exist." This would be a self-contradictory claim. Let us see why. Assume that this proposition is true. Then, given our semantical setup, we have that the item or entity in the domain, which is named by ⌜ p ⌝, has the attribute of non-existence. It is like saying of something that it both exists, since it is the referent of the constant, and yet we also claim of it that it does not exist! What has gone wrong? The answer is this: *existence is treated in our semantics like membership in a set (the domain of the model.)* This is not intuitive; it can be regarded as the price we pay for the sake of a well-behaved formalism but it can also be defended philosophically – as well as attacked on different philosophical grounds. If we treated existence as a predicate, then we would be enlisting the machinery of set theory – because, as we have seen, the meanings of predicate letters are the sets of the objects that have the predicate-property. This is not the case with existence, however. In set theory too existence cannot be contemplated as a set: for instance, we cannot meaningfully claim that we define a set $E = \{x/ x \in E\}$; this is trivially the case of any set and, hence, we have not succeeded in defining E by abstraction. This is because we must treat existence as a matter of membership only. Similarly, existence of items in the domain of a predicate model is a trivial matter. All domain objects are considered existing by construction. Accordingly, it is self-contradictory to claim of any domain member that it does not exist. (It is like saying of some member of a set that it is not a member of a set.)

- Whether the notion of an empty universe is meaningful at all has exercised philosophers for eons but it has also had ridicule heaped on it from proponents of the school of philosophy known as Analytic. It is easy to run into confusion on this score. It is an empirical matter whether something exists or not and this seems to apply to the speculative case of anything existing. Conclusions can be drawn with logical rigor from making such assumptions but the assumptions themselves have a decidedly empirical character to them and, as such, they should not be the business of logical analysis. This, of course, cuts both ways. The stipulation that bans empty membership can itself be criticized as smuggling a logical notion when this is not the case. In other words, the condition that the domain be non-empty treats this extra-logical matter as if it were logical. This has significant consequences because it affects the relationship of logical consequence: Certain formulas that express logical truths when a non-empty domain is stipulated would not be logical truths otherwise because they receive a False truth value in any model that has an empty domain. Here is such a formula:

$$\forall xFx \supset \exists yFy$$

The antecedent is true: this may seem odd but, in the metalanguage, we read the antecedent as "if there is any constant such that the constant denotes anything, then what is denoted is F." Because there can be no meaningful constants (since the constants receive their logical meanings extensionally, by what they refer to), it follows that the antecedent of this metalinguistic sentence is false and, thus, for the classical definition of implication, the whole proposition is true. Thus, in the empty domain, for any property, everything (when nothing exists) has that property. The consequent of the formula above, on the other hand, is false because it is read metalinguistically as "there is some constant, such it denotes an object and the object so denoted has the property F." Thus, with true antecedent and false consequent, the above formula is false. Yet, this is a logical truth for the standard predicate logic which excludes, by stipulation, the case of the empty domain.

Far-reaching philosophic implications have been attached to this subject. It has been claimed that this is something of a discovery that vindicates a philosophic insight made by the famous thinker Immanuel Kant who claimed that existence is not and should not be treated as a logical predicate. The stakes are quite high - if you are wondering - because many controversial philosophic proofs in the history of human thought have treated existence as if it were a characteristic; moreover, some of the most intractable philosophic problems and some of the strangest theories seem to revolve, time and again, around the issue of existence and non-existence of things. The predicate logic we are studying, instead, wipes off such problems, as it were; existence is now simply a matter of contextually specified membership in the domain of a model. Whether this is a deep insight is dubious, however. It rather seems that this situation is a direct consequence of the way we have chosen to set up our semantics in structuring a formal language in the standard predicate logic.

Here is a way in which we could indicate that something does not exist. We will see, however, that this option is not available given our semantics. We could say of an object, denoted by "a", that there is no object in our domain with which the a-denoted object is identical.

$$\sim \exists x(a = x)$$

This is logically equivalent with:

$$\forall x \sim (a = x)$$

which is, given our permitted notational conventions:

$$\forall x(a \neq x)$$

But notice what happens next. The universal quantifier gives permission, by definition, that we substitute any object whatsoever. We can then substitute a. This is because a too must refer to something. This is another semantic restriction we have - to which we turn next. Moreover, we don't have alternative domains - such as a domain restricted to non-actual or fictive entities within which, and only within which, we may allow a to refer to something. All the referents are rather restricted to our given domain D.
Making the substitution, we have:

$$a \neq a$$

Thus, the claim

$$\sim \exists x(a = x)$$

has been reduced to logical absurdity. Therefore, the following is a logical truth:

$$\exists x(a = x)$$

And, since the referent of a may well be any item in the domain, we can generalize:

$$\forall y \exists x(y = x)$$

It may seem odd that this a logical truth. Intuitively, it does not pass the test. Why should "every name has an existent object that it names" be a logical truth? This is rather the result of the choice of semantics imposed for the standard system of predicate logic. Existence is membership - and, another restriction we used and elaborated further below, there are no names that fail to refer to an element of the domain.

- We do not allow any individual constant to have no domain item to which it refers. And we do not allow ourselves to have any domain item without at least on constant naming it.

In other words, all our individual constants have values - each has one item, and exactly one item, to which it refers. And all our domain objects are labeled or tracked - each has at least one constant naming it. There are no non-referring (or uninterpreted or value-less) constants, and there are no un-named objects of the domain. We have already seen consequences that follow from this in conjunction with other restrictions on our semantics. There is another consequence that is discussed at length in logic. This is the subject of definite descriptions.

A definite description is a phrase that uniquely names something: the such-and-such. For instance, "the Queen of England," "the Eifel Tower," "the Statue of Liberty," "the President of the United States," etc... Let us look back into our stock of symbols and see that we have individual variables and individual constants: these two kinds are called TERMS. Intuitively, and also based on how we experience linguistic usage, we should expect that definite descriptions are terms. Thus, we might expect translations to go like this (if we add definite description symbols to our term symbols, stipulating that definite description symbols are "dx", where "x" is a constant letter specified in the lexicon.)

The President of the United States is a woman: W^1dp

The Queen of England visited the Eifel Tower: V^2dqde

The Statue of Liberty is in New Jersey: N^1dl

Nevertheless, this is not to happen. The reason is that we would run into difficulties - indeed, into absurdity - with non-denoting or non-referring definite descriptions. Let us use an example that has been memorialized over countless renditions of logic textbooks. Suppose that we symbolize "the present King of France" by "dk." We examine now the proposition made by the sentence,

The present King of France is bald.

Naively continuing with our assumed rendering of definite descriptions as terms, we have the following translation:

Bdk

Let us take the domain of all presently actually existing persons. We cannot place some person who is allegedly the King of France in the domain. This presents the unhappy prospect that we cannot consider our proposition to be the kind of thing that is either true or false. "dk" has nothing to refer to - and this is not allowed. Of course, we do not allow ourselves to deal with propositions that might be neither true nor false (which is available as a foundational choice for constructing

non-classical or alternate logics.) A solution that occurred to Gottlob Frege, the father of modern logic, but lapsed into oblivion, is to allow for a non-object as being a member of our domain; in that case, we can treat this non-object as the referent of all non-referring terms. In that case, we could stipulate that "Ft" is false whenever the value or referent of t is the non-object of our domain. But this solution is not available.
Alternatively, we could declare ourselves liberated from empirical and historical considerations and pretend that the entity "the present King of France" is in our domain. This, however, makes this entity automatically existent - for reasons we have already recounted. This is not an acceptable solution to our predicament, either.

The standard solution, counterintuitive as it is, owes to one of the most prolific writers in many subjects, including Logic, Bertrand Russell. The solution is called regimentation. Here is what this means. Regardless of what ordinary language does, we regiment when we translate into our formal predicate language in such a way that we do not respect the usual linguistic practice; but in so doing, we claim that it is our formal language - and not ordinary language - that gets it right. We claim to be eliciting or bringing out the proper logical grammar of a claim and this logical grammar has been basically distorted and missed by the habitual user of ordinary language - even if this user is linguistically competent but still innocent of what the underlying logical grammar of a language is.

The regimentation for definite descriptions goes like this. First, definite descriptions are not treated like names or terms. The whole phrase that has the definite description in it is spelled out in a certain way. To begin with:

(S1) The present King of France is bald

does not become,

(S2) Bald (the-present-King-of-France)

where "the-present-King-of-France" is obviously a term. Even though this seems to be the standard linguistic practice with respect to definite descriptions, instead, our Russellian regimentation proceeds as follows:

There is a unique x, such that x is the present King of France and x is bald.

We notice that "being-King" must be treated as a logical predicate now. What happens to the truth value of the above proposition? We are saying that there is no entity in our domain such that is in the extension of the predicate King-of-France and is also in the extension of "bald." Thus, we must assign in the model the truth value False to the proposition.

We could even start with the claim:

(S3) There is a unique x, such that x is the-present-King-of-France.

In light of what we have said above about how existence is treated in our semantics, we have the following. Notice that we have no term "the-present-King-of-France", which would pose problems of reference in our domain. Instead, we have an empty logical predicate (the-present-King-of-France), which is allowed. Thus, we cannot say of any constant that it names some object in the extension of the-present-King-of-France: this extension is specified as the empty set in our domain. Hence, the above proposition is false. Similarly, the proposition,

(S4) There is a unique x, such that x is the present King of France and x is bald.

will also turn out to be false. (We will learn in due time how to translate uniqueness claims. We will do this with the help of identity, for which we already have a symbol in our grammar and we have stipulated semantics.)

This may be thought of as counterintuitive too. Why is the claim "the present King of France is bald" false if no such person exists? Are we talking about his existence or about his being bald? If the latter, it may seem like a more intuitively appealing response to say that the claim cannot be evaluated - it is to be considered as neither true nor false.

This characterization may be a matter of failure to assign a truth value or, controversially, it might compel us to the adoption of a non-standard logical formalism that includes a special truth value for “neither true nor false.” But, as in standard Propositional Logic so in standard predicate logic, we must treat no meaningful propositions as being neither true nor false. (And one may argue, indeed, that our sentence comes across as expressing a meaningful proposition, regardless of the difficulties we are having with it.)

Moreover, should we take the proposition made by the sentence "the present King of France is *not* bald" to be true as being the negation of the false proposition of "the present King of France is bald"? Or should we take as being false since, in this case too, no such person exists?
Russell had an answer to this quandary, eventually, and we will turn to that briefly after we have learned how to symbolize or translate uniqueness claims. (Remember that definite descriptions have a uniqueness element to them. "The President of the United States", rendered in regimentation, must then be "the unique x, such that x is the unique member of the extension of "the President of the United States." In the language of set theory, the extension of the predicate "the President of the United States" is a singleton: a set with exactly one member. We don't know yet how to symbolize "there is a unique x, such that x is a member of set X." But we will find out how subsequently.)

Having honed our appetites, we will be looking forward to symbolization or translation (from English) into our formal language; taking advantage of the fact that we have symbolic resources, and formal provisions, for identity, we will be able to symbolize this Russellian regimented phrase, "there is a unique x such that x Fs," which is said to serve as our correct rendering of a definite description ("*the*-F."). No special quantifier symbol is needed for the semantic accommodation of “there is a unique x” – although we could add it and define it in terms of the universal and existential quantifiers (which are also inter-definable as we know.)

We could add a uniqueness quantifier, symbolized by "∃!" and grammatically accompanied by an individual variable letter, but this quantifier is definable in terms of the quantifier symbols we already have. We can give the following two logically equivalent definitions and also show how the proposition of the famous sentence "the present King of France is bald" (KFrB) is to be translated in regimentation with predicate letters "K" and "B".

(∃!df1) $\exists!xFx \stackrel{def}{=} \exists x(Fx \cdot \forall y(Fy \supset (x = y)))$
(∃!df2) $\exists!xFx \stackrel{def}{=} \exists x\forall y(Fy \equiv (x = y))$
(KFrB1) $\exists x(Kx \cdot \forall y(Ky \supset (x = y)) \cdot Bx)$
(KFrB2) $\exists x\forall y((Ky \equiv (x = y)) \cdot Bx)$

❖ Expansions and Interchanges of Quantifier Phrases

The two quantifiers we have in our predicate logic can be defined provisionally in terms of standard logical connectives of Propositional Logic. This is allowed only if we deal only with finite domains. When we limit ourselves to the examination of specified finite models, we can reap certain benefits from defining the universal quantifier in terms of conjunction and the existential quantifier in terms of inclusive disjunction. Of course, we are familiar with the logical connectives of conjunction and inclusive disjunction from the study of Propositional Logic.

Let us consider universal quantification first.

$\forall xFx$

This means that all objects in the given domain of a model are in the extension of the logical predicate, also characterized in the model, which is symbolized by "F." Assume that we have domain members as follows:

$D = \{o_1, o_2, ..., o_n\}$

In other words we have n objects in the domain, with the "o_j" symbolization for the objects and j a member of the subset of natural numbers {1, 2, ..., n}.

Of course, we allow ourselves a stock of individual constants as needed; this is given us by the grammatical arrangements we have in predicate logic. The values of the individual constants are:

$\surd_M(a_1) = o_1$
$\surd M(a_2) = o_2$
...
$\surd_M(a_n) = o_n$

For the universally quantified phrase we have above to be true, it must be the case all objects of the domain are in the extension of F. We know all this from our semantical arrangements. It must then be true for every individual constant, a_j:

$\surd_M(Fa_j) = T$

This is because the referent or value of the constant is in the extension of F.

Given the definition of conjunction, we then have:

S ince $\surd_M(Fa_1) = \surd_M(Fa_2) = ... = \surd_M(Fa_n) = T$, then $\surd_M(Fa_1 \cdot Fa_2 \cdot ... \cdot Fa_n) = T$

This means that,

Given that $\surd_M(\forall xFx) = T$, then $\surd_M(Fa_n) = T$, then $\surd_M(Fa_1 \cdot Fa_2 \cdot ... \cdot Fa_n) = T$

We can also prove the converse:

Given that
$\surd_M(Fa_n) = T$, then $\surd_M(Fa_1 \cdot Fa_2 \cdot ... \cdot Fa_n) = T$, then
$\surd_M(\forall xFx) = T$

This is obvious intuitively and it follows trivially from the definitions of the various concepts. Since every object in the domain is in the extension of F, then the truth value of the quantificational phrase "all things in the domain are F" must be true.

If you want to be a stickler for detail you should notice that something is actually amiss in the step of proving the converse:

Given that $\surd_M(Fa_n) = T$, then $\surd_M(Fa_1 \cdot Fa_2 \cdot ... \cdot Fa_n) = T$, then
$\surd_M(\forall xFx) = T$

Here we need the additional assumption, which we made, that the constants $\{a_1, a_2, ..., a_n\}$ exhaust referentially all the elements of the domain. This extra assumption breaks the symmetry between ⌜ $\forall xFx$ ⌝

and⌜ $Fa_1 \cdot Fa_2 \cdot ... \cdot Fa_n$ ⌝ and is the reason why, strictly speaking, we cannot define universal quantification as a conjunction. To have a definition we must have logical equivalence obtain between the two related expressions. In this case we would have to have:

$$\forall xFx \equiv (Fa_1 \cdot Fa_2 \cdot ... \cdot Fa_n)$$

Even though we admit that, given the possibility of an infinite domain. the universal quantifier cannot be precisely defined as conjunction (contrary to a bright idea that originally goes back probably to Ludwig Wittgenstein), we will, nevertheless, avail ourselves conveniently of a license to treat universal quantification as if the above logical equivalence obtained. We call this the development of the universal quantifier. Given a model M, the universal quantifier quantifies over the elements of the model's domain, and we take its expansion or development to be as follows for a unary or monadic predicate symbolized by "F."

$$\partial(\forall xF^1{}_j x) = F^1{}_j a_1 \cdot ... \cdot F^1{}_j a_n$$

when the number of elements in the domain is n.

In the case of the existential quantifier, and given the definition of this type of standard predicate logic quantifier, we can also settle on an expansion formula. Recall that the existential quantificational phrase means, for some monadic predicate constant symbolized by "F", that "at least one element of the domain - we know not which one - is in the extension of the predicate F." This matches the classical connective we have called inclusive disjunction.

$\surd(p_1 \vee ... \vee p_m) = T$ if and only if (iff) at least one of $\surd(p_1), ..., \surd(p_m)$ is T

The expansion for the existential quantifier, then, must be as follows:

$\partial(\exists xF^1{}_i x) = F^1{}_i a_1 \vee ... \vee F^1{}_i a_m$, when the number of members of the domain (the cardinality of the domain) is exactly m

We can extend these notions to n-ary logical predicates. Consider the case of a binary logical predicate (or binary relation).

$\partial(\forall x_1 \forall x_2 F^2{}_j x_1 x_2) = F^2{}_j a_1 a_1 \cdot F^2{}_j a_1 a_2 \cdot F^2{}_j a_1 a_3 \cdot ... \cdot F^2{}_j a_2 a_1 \cdot$
$F^2{}_j a_2 a_2 \cdot ... \cdot$
$F^2{}_j a_{n-1} a_n \cdot F^2{}_j a_n a_n$,
when the number of elements in the domain is n.

The expansions we have introduced in this section are truth-functional. There is no prospect, however, of somehow using such tricks to end up with mechanical procedures, as in Propositional Logic. Ultimately, we cannot circumvent the complexity of predicate formulas (stemming from the availability of resources for analyzing the internal structure of propositions). It was proven by Alonzo Church, in 1936, that there is no effective mechanical decision procedure (like the truth table in Propositional Logic), which can be applied to yield correct results always in the case of predicate or quantificational logic. The expansions we examine in this section only have localized significance and convenience working with specified models. In the next section, we will investigate the subject of validity of argument forms in predicate logic. In preparation, we can say this. If a model can be supplied in which the premises of an argument form are true and its conclusion is false, this has effectively shown that the contemplated argument form is invalid. Even one counterexample - or, as we should say here, countermodel - suffices to invalidate a suggested argument form. Nevertheless, the discovery of a countermodel, if one is available, is not a mechanical test. We may or may not discover a countermodel, even if one is available, as we search for it. We have no recipe for proceeding with model construction in such a way that application of mechanical instruction is guaranteed to yield a countermodel if and only if such one exists.

I.6 EXAMPLES

- We examine an example of writing out expansions of quantificational formulas within a given model.

M = <D, √>
D = {♚, ♛, ♜}
$\surd_{M}(a)$ = ♚
$\surd_{M}(b)$ = ♛
$\surd_{M}(c)$ = ♜
$\surd_{M}(F)$ = {♚, ♛}
$\surd_{M}(G)$ = {♚, ♛, ♜}

$\surd_{M}(R) = \{<$♚, ♛$>, <$♚, ♜$>, <$♛, ♜$>\}$
$\surd_{M}(f(a)) =$ ♛ $= \surd_{M}(b)$
$\surd_{M}(f(b)) =$ ♚ $= \surd_{M}(a)$

We could choose to tell a more concrete story by giving linguistic interpretations like the following. Nevertheless, the model would check exactly the same formulas as true and false regardless of adding this elaborate linguistic gloss.

INT(Fx): x is a royal persona
INT(Gx): x is a player in an elaborate power-game
INT(Rxy): x prevails over y in confrontation
INT(f(x)) = spouse of x

Now we determine the truth values in the model of the following propositional formulas. First, we try to acquire the habit of assessing truth or falsehood by inspecting the model. This might not always be possible depending on the complexity of the formulas given.

The developments we have learned in this section come handy in this respect, as we will see. The truth value of quantificational formula has to be the same as the truth value of its expansion. We take advantage of this.

$$\surd_{M}(\partial(\forall x Fx)) = \surd_{M}(Fa \cdot Fb \cdot Fc))$$

Now we check our model. The referent (denotation, interpretation, value) of the constant a is in the extension (interpretation, value) of F. The same is true of the value of b but not of the value of c. Observe now how we will use some basic set-theoretical notation (which you had to review prior to embarking on the study of predicate logic). We use this notation in our metalanguage (this mix of English and symbols we use to talk about our formal system.)

$$\surd_{M}(a) \in \surd_{M}(F) => \surd_{M}(Fa) = T$$

[This says: the value of a is a member of the value of F. In more intuitively appealing speech: the referent of a, what a names or labels, is an object that has the F-property. Therefore, the proposition expressed by Fa is true or the truth value of Fa is T.]

$$\surd_{M}(b) \in \surd_{M}(F) => \surd_{M}(Fb) = T$$
$$\surd_{M}(c) \notin \surd_{M}(F) => \surd_{M}(Fc) = F$$

In this case: the value or referent of c is not a member of the extension or value of F. Therefore, the truth value of Fc is F or false. Next, we simply continue with our determination of truth value knowing what the definition of conjunction is.

$\surd_M(Fa \cdot Fb \cdot Fc) = T \cdot T \cdot F = T \cdot F = F$

But we know that the truth value of the universal quantificational phrase is the same as the truth value of its development. Thus, we have determined the truth value *in the given model* of the proposition symbolized by "∀xFx".

$\surd_M(\forall xFx) = \surd_M(\partial(\forall xFx)) = F$

We can proceed similarly with what become computations of truth values for other quantificational propositions. It is a good idea too to examine a model informally and try to determine the truth value of propositions in the model without necessarily going through the travails of expansions. For instance, one can immediately examine the given model above and diagnose that in this model the truth value of ∀xFx is false since not all items in the model's domain are F. This approach becomes untenable, though, when we deal with complex formulas.

$\surd_M(\forall xGx) = \surd_M(\partial(\forall xGx)) = \surd_M(Ga \cdot Gb \cdot Gc) =$
$T \cdot T \cdot T = T \cdot T = T$

This is so given that the values of all the constants (all objects in the domain) are in the extension of G.

$\surd_M((\exists xFx \cdot \exists xGx) \supset \exists x(Fx \cdot Gx))$

Let us approach this implicational propositional formula in a different way. We may recall that standard Propositional Logic has a certain interesting characteristic, spelled out below. This characteristic is preserved in predicate logic because the definitions of logical connectives are still the same.

$\varphi_1, \varphi_2, \ldots \varphi_n \vDash \psi$ if and only if $\vDash (\varphi_1 \cdot \varphi_2 \cdot \ldots \cdot \varphi_n) \supset \psi$

If the logical form with premises $\varphi_1, \varphi_2, \ldots, \varphi_n$ and conclusion ψ is valid, then the implicational proposition with antecedent the conjunction of the premises $\varphi_1, \varphi_2, \ldots, \varphi_n$ and consequent the conclusion ψ.

$(\varphi_1 \cdot \varphi_2 \cdot \ldots \cdot \varphi_n) \supset \psi]$

is necessarily true or cannot be falsified.

This relationship between logical validity (logical consequence, validity of argument forms) and implicational propositions applies in the case of examining truth value computations within any model of predicate logic.

So, to examine if the given implicational propositional formula is true in the given model M, we might as well examine if the following argument form is valid in the model:

$$\exists xFx \cdot \exists xGx \vDash ?\ \exists x(Fx \cdot Gx)$$

Quick inspection shows that this is the case. There is at least one F and there is also at least one G. Let us be careful about how we proceed. We attempt to falsify the proposed conclusion - that there is an object that is both F and G. We should get that there is no such object. But this is not the case since the objects denoted or referred to by a and b are both F and G.

We can also apply the expansion method we have learned in this section.

$$\surd_M(\exists xFx \cdot \exists xGx) = \surd_M(\partial(\exists xFx \cdot \exists xGx)) = \surd_M(\partial(\exists xFx) \cdot \partial(\exists xGx)) = \surd_M((Fa \vee Fb \vee Fc) \cdot (Ga \vee Gb \vee Gc)) = (T \vee T \vee F) \cdot (T \vee T \vee T) = T \cdot T = T$$

Now, we must falsify the given conclusion to see if that is feasible. If so, we have invalidity for the given argument form *in the model*. Otherwise, the formula is valid *in the model*.

We pretend: $\surd_M(\exists x(Fx \cdot Gx)) = F$

Thus, we have:

$$\surd_M(\exists x(Fx \cdot Gx)) = \surd_M(\partial(\exists x(Fx \cdot Gx))) = \surd_M((Fa \cdot Ga) \vee (Fb \cdot Gb) \vee (Fc \cdot Gc)) =$$
$$= (T \cdot T) \vee (T \cdot T) \vee (F \cdot T) = T \vee T \vee F = T \vee F = T$$

But we have been assuming that,

$$\surd_M(\exists x(Fx \cdot Gx)) = F$$

So, we have reached a contradiction. It is logically impossible to invalidate the given argument form in this particular model. (Do not assume, though, that this is a valid propositional argument form. The form has been shown valid only in one particular model. This particular model is not a countermodel - or counterexample to the argument form. But this does not show that there is *no* countermodel whatsoever.)

$$\surd_{\mathcal{M}}(\exists x(x = f(a))) = \surd_{\mathcal{M}}(\partial(\exists x(x = f(a)))) = \surd_{\mathcal{M}}((a = f(a)) \vee (b = f(a)) \vee (c = f(a))) = \surd_{\mathcal{M}}(F \vee T \vee F) = T$$

$$\surd_{\mathcal{M}}(\forall x(x = f(c))) = \surd_{\mathcal{M}}(\partial(\forall x(x = f(c)))) = \surd_{\mathcal{M}}((a = f(c)) \cdot (b = f(c)) \cdot (c = f(c))) = \surd_{\mathcal{M}}(F \cdot F \cdot F) = F$$

Matters become somewhat more complicated with respect to developing the expansions, when we deal with many-place predicates and with iterated quantifiers or combinations (also called nestings) of quantifiers. We will examine relevant examples now.

$\surd_{M}(\forall xRxx) = \surd_{M}(Raa \cdot Rbb \cdot Rcc) = F \cdot F \cdot F = F \cdot F = F$

Indeed, immediate inspection too could have shown that this is the case. Nothing is R-related to itself: of no object or entity in the domain is it said for the given model that it prevails over itself. We should not read deeper implications into this; it's just that this is the setup or construction of the given model. Analytically, we have:

$<\surd_{M}(a), \surd_{M}(a)> \notin \surd_{M}(R)$

$<\surd_{M}(b), \surd_{M}(b)> \notin \surd_{M}(R)$

$<\surd_{M}(c), \surd_{M}(c)> \notin \surd_{M}(R)$

As we should know, extensions of two-place or binary predicates are represented as sets of ordered pairs. The objects in the ordered pair $<x, y>$ are so arranged that x is R-related to y. We are not told anything about whether y is R-related to x. For that we would have to have $<y, x>$ also in the extension of R.

The objects x, y, ... are the values of the constants in the system. Thus, for instance, since the item of our domain symbolized by "♛" is not R-related in our model, we have that the ordered pair $<\text{♛}, \text{♛}>$ is not in the extension or value of R:

$$<\text{♛}, \text{♛}> \notin \surd_{M}(R)$$

Given that

$$\surd_{M}(b) = \text{♛}$$

we have:

$$<\surd_M(b), \surd_M(b)> \notin \surd_M(R)$$

- Here is another example.
  $\surd_M(\exists xRxx) = \surd_M(\partial(\exists xRxx)) = \surd_M(Raa \vee Rbb \vee Rcc) = F \vee F \vee F = F \vee F = F$

Next, we consider nesting of quantifiers. Let us take the first example.

$$\surd_M(\forall x\exists yRxy) = ?$$

We must end up with an expansion that is conjunctional. This is because the quantifier of the larger scope is the universal quantifier. Notice how we proceed analytically to take the expansion of this propositional formula. We start from expanding the lesser-scope quantifier first and the larger-scope quantifier next. We note that, in this formula, the lesser-scope quantifier is existential - hence compelling us to take an inclusive disjunction - and the larger-scope quantifier is universal - hence compelling us to take a conjunction.

$$\surd_M(\forall x\exists yRyx) = \surd_M(\partial(\forall x\exists yRyx)) =$$
$$\surd_M(\forall x(Rax \vee Rbx \vee Rcx)) =$$
$$(Raa \vee Rba \vee Rca) \cdot (Rab \vee Rbb \vee Rcb) \cdot (Rac \vee Rbc \vee Rcc) =$$
$$(F \vee F \vee F) \cdot (T \vee F \vee F) \cdot (T \vee T \vee F) = F \cdot T \cdot T = F \cdot T = F$$

Examining the model, we should expect this result. Reading the given formula - or translating into English - we have: all entities are dominated by at least one entity. Analytically,

all x are such that there is at least one y such that y dominates x

This is not true. The item referred to by a is not dominated by any entity - including its self.

- Some observations are in order in anticipation of more on this subject, which follows in the next section. We have established that the truth value of ⌜ ∀xFx ⌝ is false in this model. Since we have shown a falsehood, we have a falsifying model to the propositional formula ⌜ ∀xFx ⌝. Our model is one (of an infinite number of) countermodels to a suggested claim that ⌜ $\forall xFx$ ⌝ is a predicate tautology or logical truth. We can write:

$$M \nvDash \forall xFx$$

This suffices as a countermodel. It is hardly surprising that the claim that everything has a property, let's say F, is *not* a logical truth. No such claim about any property should constitute a truth of deductive logic. It is an empirical matter rather than a matter of truth conditions (definitions of logical connectives and other logical phrases, like "all" and "some") that things do or do not have some property. Even if, by some fluke, all things happen to have a property in some world - even in the one we consider our one and only actual world - we still don't want to consider this as a matter of logic. There should be always a countermodel we can construct to the claim

$$M \vDash \forall xSx$$

for any S.

In the model we have examined, we established the following:

$$M \vDash \forall xGx$$

This surely doesn't allow us to claim that the formula ⌜ $\forall xGx$ ⌝ symbolizes a logical truth. The same considerations we brought up in the preceding paragraph apply here. It cannot be a matter of logic - and so, it cannot be a logical truth - that items possess some property or not. So, apparently, the verification of the claim ⌜ $\forall xGx$ ⌝ in a model should not be taken as establishing that this is a logical truth. This is not hard to see. After all, we could easily construct another model, M', in which ⌜ $\forall xGx$ ⌝ is false. But a logical truth should be true in every model! Because a logical truth does not depend on empirically verifiable or non-logical considerations, its status is independent of the information that we put into models through the extensions of the predicates. Now you can understand why logical predicates are sometimes called non-logical constants. They are fixed but only per each model. (We make only one exception. This has been already mentioned. Recall that the only exception we make is for the identity relation.)

The quantifier phrases we have in our semantics are inter-related. Let us be clear that we only have two quantifiers: the universal quantifier and the existential quantifier. This is how they should be understood in their semantics:

$\forall xFx$: all objects in the given domain have the F-property (are in the set that is the value or extension of F in the given domain.)

$\exists xFx$: at least one object in the given domain has the F-property (is in the set that is the value or extension of F in the given domain.)

We have no expressive resources and no way of semantically accounting for such quantificational phrases of English as:
most x F; few x F; exactly half of the given things F; etc....

There are ways to provide semantical arrangements that can accommodate such expressions. If you think hard on this subject, you might figure out that we could use the cardinal numbers of our domain and of the extensions of predicates, and then make certain decision as how "few" or "most" and such phrases are to be calibrated quantitatively. We will not pursue this here since it is not part of the standard approach to the study of predicate logic.

It is imperative to learn the relationships between various quantificational phrases. Interestingly, negations can be entered into interplay with quantification in two ways. The negation can be applied to the whole quantificational phrase; or the negation can follow the quantificational phrase. In other words, the negation can be applied to the quantifier or be within the scope of the quantifier. Here are all the possibilities:

$\sim \forall xFx, \forall x \sim Fx, \sim \exists xFx, \exists x \sim Fx$

We will now express the relations - also called interchanges - between quantificational phrases by means of logical equivalences. As we know from the study of Propositional Logic, two propositions are logically equivalent if and only if they have exactly the same truth values in all cases - which means for all valuations. In Propositional Logic, a case or valuation is an assignment of truth values to the components of a proposition; we can think of a valuation as a row of the truth table if we are using that approach to the semantics. In predicate logic, we have already discerned that valuations apply to internal parts of sentences - like predicate phrases and names and quantificational phrases. A valuation requires a model. We can say that the equivalences we give below hold in every logically possible or constructible model. They are indeed logical truths.

$$\forall xFx \equiv \sim \exists x \sim Fx$$
$$\sim \forall xFx \equiv \exists x \sim Fx$$
$$\forall x \sim Fx \equiv \sim \exists xFx$$
$$\sim \forall x \sim Fx \equiv \exists xFx$$

One could actually use only one of the two quantifiers and define the other one derivatively. For instance, we could include only the universal quantifier in our formal setup and then permit that the existential quantifier can be defined in terms of the universal:

$$\exists xFx \overset{def}{\equiv} \sim \forall x \sim Fx$$

We would not have bought into any anomalous consequences because, indeed, this is the correct relationship between the two quantifiers of standard predicate logic.

First let us make intuitive sense of the above interchanges. Then we can also use what we have laid down about quantifier expansions in this section to confirm the results. (Recall, however, that taking advantage of expansions would not be permitted in more formal or thorough investigations.)

$$\forall xFx \equiv \sim \exists x \sim Fx$$

all objects of the model have the F-property if and only if no object of the model fails to have the F-property;

all members of the domain are in the extension of F if and only if no member of the domain is not in the extension of F.

$$\sim \forall xFx \equiv \exists x \sim Fx$$

not all objects of the model have the F-property if and only if at least one object of the model fails to have the F-property;

not all members of the domain are in the extension of F if and only if at least one member of the domain is not in the extension of F.

$$\forall x \sim Fx \equiv \sim \exists xFx$$

all objects of the model fail to have the F-property if and only if no object of the model has the F-property;
all members of the domain are not in the extension of F if and only if no member of the domain is in the extension of F.

$$\sim \forall x \sim Fx \equiv \exists xFx$$

not all objects of the model fail to have the F-property if and only if at least one object of the model has the F-property;
not all members of the domain are not in the extension of F if and only if at least one member of the domain is in the extension of F.

Here we have occasion to celebrate the contribution our formal language makes to removal of ambiguity. Let us take a sentence of everyday language like,

(S) Everyone did not come.

This is an ambiguous sentence. It could be expressing the proposition, "it is not the case that everyone came" or it could be expressing the proposition "it is the case about everyone that he or she did not come." This is an ambiguity about the scopes of negation and of the quantifier. Is the quantifier within the scope of the negation or is the negation within the scope of the quantifier. We disambiguate by providing both options without taking a stance as to which one is intended since we do not have any additional information about the context of the linguistic utterance. The disambiguating propositions (pdS) are as follows (with lexicon given as "Cx" stands for "x came."):

(p1dS) $\sim \forall xCx$

(p2dS) $\forall x \sim Cx$

We can use expansions to confirm the relations between quantifier phrases. Let us use a model constructed over a domain with two objects (with the case easily generalized to models with n-numbered domains.)

- We can use predicate-extensions diagrams to fix the characteristic signature of a given model $\mathcal{M}$. An example follow. In the Appendix I on Set Theory we also show how we can apply set-theoretic tools to the sets that are model-theoretically assigned as extensions of predicate symbols. By doing this we can present interesting results systematically. Standard disjunction corresponds to– or is set-theoretically interpreted by – union of sets; conjunction to overlapping or intersection; equivalence to equality of sets; and implication to the union of the complement of the antecedent set and the consequent set. But, for now, we have an example in which complete information about the signature of the model utilizes predicate-extension diagrams. Inclusion in the extension is symbolized by "+" and failure of exclusion by "–". For the binary relation, the horizontal axis is considered as listing the first members of the ordered pairs whose set defines the binary relation.

D = {①, ②, ③}
√(a) = ①
√(b) = ②
√(c) = ③

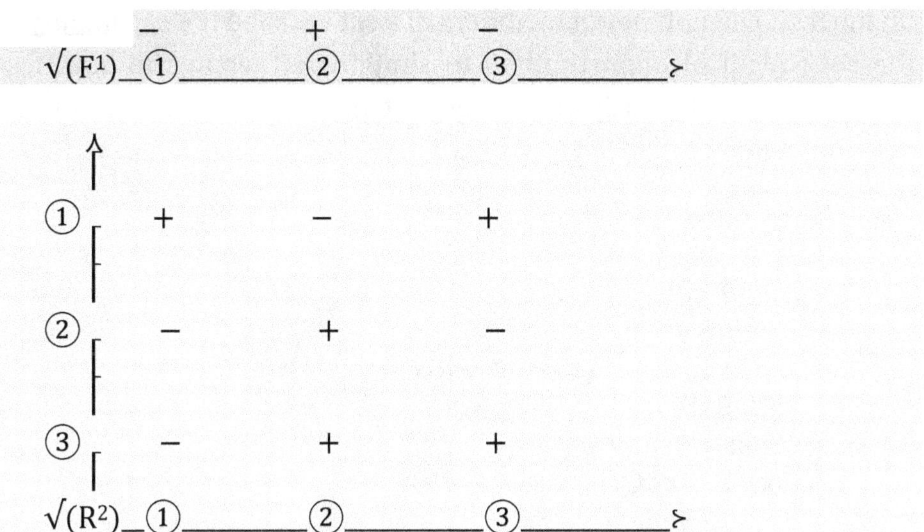

We can extract the values from the given diagrams.

$\surd(F^1) = \{b\}$
$\surd(R^2) = \{<a, a>, <c, a>, <b, b>, <b, c>, <c, c>\}$

$\surd(F^1c) = F$
$\surd(R^2ac) = F$
$\surd(\exists yR^2by) = T$
$\surd(\forall x\forall y)(F^1x \supset F^1y) = F$

[For instance, $\surd(F^1b \supset F^1a) = F$; thus at least one of the conjuncts in the development of the formula is false.]

$\surd(\forall x\forall y\exists z(Rxy \equiv (z \neq x \cdot z \neq y \cdot F^1z))) = F$

[<b, b> is in R but there is no constant in F that denotes a distinct object from that denoted by b.]

I.7 EXERCISES

(1) Write out the developments of all the following well-formed formulas (wffs) of PRED for models with domains of two and three members. (The number of members of a set is called the cardinality of the set.) Avail of opportunities to simplify, either at the outset, if possible, or after you have carried out the mandated steps for the development.

a. $\forall x\forall yRxy$
b. $\exists x\exists yRxy$
c. $\forall x\exists yRxy$
d. $\exists x\forall yRxy$
e. $\forall x(Fx \supset Gx) \supset (\forall xFx \supset \forall xGx)$
f. $Fa \supset \exists xFx$
g. $\forall xFx \equiv \exists xRax$
h. $\exists x(Fx \supset \forall yGy)$
i. $\forall yGy \supset \exists xRax$
k. $\exists xFx \cdot \exists yFy$
l. $\forall z\forall w((z = w) \vee Rwz)$

m. ∃x((f(x) = b(x)) ≡ Sf(x)m(x))
n. ∃x(Fx · Gx) ≡ (∃xFx · ∃xGx)
o. ∀x(Fx ∨ Gx) ≡ (∀xFx ∨ ∀xGx)

(2) Determine the truth values of the given wffs in the specified models.

a.
D = {⨹, ⨺, ⨻}
√(a) = ⨹
√(b) = ⨺
√(c) = ⨻
√(F) = {⨹, ⨻}
√(R) = {<⨺, ⨺>, <⨺, ⨹>, <⨺, ⨻>}
∃x∀y(y ≠ x ⊃ (Rxy · ~ Ryx))
∃x(∀yFy ⊃ Rxy)
∃x∃y(Rxy ∨ Ryx)
∀x∀y(Rxy ≡ Ryy)
Rab ⊃ ∀x∀y∀z(Rxy · (Ryz ≡ Rxz))
∃x∀y(Rxy ≡ (Fx · Fy)) ⊃ Rbb

b.
D = {⊙, ⊕, ⊗}
√(a) = ⊙
√(b) = ⊗
√(c) = ⊕
√(R) = {<⊙, ⊗, ⊕>, <⊗, ⊕, ⊙>}
√(F) = {⊗}
√(G) = {⊙, ⊕}
∃x∃y(x ≠ y · ∃zRxzy)
(Ga · Gc) ⊃ ∃z(Razc · ∀w(Rawc ⊃ w = z))
∀x∀y(Fx ⊃ (Gy ⊃ ∃zGxzy))

Now consider the following interpretation imposed on this model.

Fx: x is a person.
Gx: x is a thing.
Rxyz: x gives y to z.

Does anything change with respect to what formulas are satisfied by the model?

Given the superimposed interpretation, consider if the propositions expressed by the following English sentences are true or false in the model. (Not having learned formal translations yet, this is an exercise in scanning the information given in the construction of the model. Is this a reliable method for determining truth values?)

a. Someone gives something to someone.
b. Not everyone gives everything to everyone.
c. Someone gives something to someone else.
d. Something is given to someone or nothing is given to anyone.
e. If someone gives something to himself, then everyone gives something to everyone.

(3) Is there any model that satisfies the following well-formed formulas? [Clues: Define logical contradictions. Also, recall that well-formed formulas of predicate logic may not be expressing propositions...]

$\forall x(Fx \equiv {\sim} Fx)$
$\exists x \forall y(Rxx \equiv {\sim} Rxy)$
$\forall x \forall y(Rxy \supset (Fa \vee {\sim} Fz))$
$\exists x Fx \equiv {\sim} \exists y Fy$
$\exists x(\exists y Fy \supset (Fxx \cdot {\sim} Fyx))$

(4) As we have indicated, we cannot translate with our symbolic resources quantifier phrases like "most," "a few," "a lot," and other such expressions. In this exercise we give a hint as to how alternative formal systems can be constructed to accommodate non-standard quantifiers. If you need, consult Appendix I on Set Theory. Study the example given below and then do the exercises.

Formal Language PREDSET
Symbols for sets: X, Y, Z, ..., X_i, ...
Domain: D
Individual variables: x, y, ..., x_i, ...

<u>Atomic constants</u>: a, b, ..., a_i, ...
x is a member of X: x ∈ X

The cardinality (card) of a set is the number of its members.

Example: D = {1, 2, 3, 4, 5}
F = {x/ x is odd} = {1, 3, 5}
G = {x/ x is even} = {2, 4}
MOSTxFx if and only if card(F)/card(D) > ½

The proposition "most numbers are odd" is true in the given model.

card(F)/card(D) = 3/5 > ½

The proposition "most numbers are even" is false in the given model.

card(G)/card(D) = 2/5 < ½

a. Assess the propositions expressed by the sentences "most X are Y," "a few X are Y," "most Y are X," and "a few Y are X" in the model:
<D = {⊏, ⊐, ⊑, ⊒}, X = {⊏, ⊒, ⊑}, Y = {⊏, ⊐}>.

[How are you going to specify the conditions for the quantifying expression "a few?"]

[Notice that there are available options when it comes to setting cardinalities ratios. This was true in the preceding example but we adopted one of several alternative, reasonable, options for defining the quantifier "most." Do you think that this availability of options – the fact that there is no fixed or rigid regulation of this subject – makes logics with non-standard quantifiers in any sense not "pure" logics?]

b. <D = {⊤, ⊥, ∓, ±}, X = {⊤, ⊥}, Y = {∓, ±}, Z = {<∓, ±>, <⊤, ∓>, <⊤, ±>}>

Assess the truth values of the following sentences' propositions in the given model. Define non-standard quantifiers in the formal language before you can proceed with the assessments.

i. More Xs are Z-related to Y than Y are Z-related to X.

ii. Some X are Y.

iii. Most Y that are Z-related to X are not Z-related to themselves.

iv. Half of Ys are Z-related to Xs.

v. No Xs are Ys.

vi. If no Xs are Z-related to themselves, then a few Ys are Z-related to themselves.

(5) Here is an informal proof. Can we translate this proof into the formal language PROP, given the symbolic resources and arrangements that are available to us? If yes, how? If not, why not? *Assume that two names, a and b, co-refer. Since they name the same entity, it follows that their referents share all their properties. One such property is necessary self-identity: the object that is the referent of both a and b is necessarily identical with itself. Let us consider this to be referent of a. We can say, then, that a is necessarily identical with itself – which is not only intuitively but means, for us, that the object referred to by a is necessarily self-identical. Since the object referred to by b has all the properties possessed by the object referred to by a, then this object is also "necessarily identical with the object referred to by a." Thus, b is necessarily identical with a. We have proved that, if any two names co-refer, then it is logically necessary that they co-refer. (An instance of this is: Morning-Star and Evening-Star happen to refer to the same celestial object – Venus. Therefore, given the proof above, it is logically necessary that they name the same object. This is a surprising result, of course, considering that it was not even known for long centuries that they named the same object. Or, does this matter at all?)*

(6) We can provide an informal proof of the following well-formed formula of PL. This can be read as "everything in the model exists." Existence, rather than being queried as a subject of arcane metaphysics, is relativized ad hoc to model inclusion. This is a subject of considerable philosophic significance. It may be objectionable but the justification that can be offered goes like this: what exists should not be a matter of abstract speculation; after all, this is an empirical matter. Whatever theory is given us – extra-logically – its model stipulates what objects exist. This is the proper way to manage mattes of existence. For instance, in science, which is well-behaved when it comes to managing speculations without running into excess, the theory specifies what kinds of things exist (for instance, homomorphic space-points in Newtonian physics or space-time in Relativistic

physics.) At any rate, the fact is that our predicate formalism has this as a logical truth.

$\vDash \forall x \exists y(x = y)$

Assume that it is not the case that, for every name x there is a name y that actually refers to an object and co-refers with x. Then, by the definitions of the quantifiers and their interchange relationship and interactions with negation, we have: there has to be a name x that refers to an object, such that for every name y, x does not co-refer with y. Call this referring name "a". Then, based on what we have just said, for all names y, a does not co-refer with y. But we can replace y by a, since we are saying that this is the case for "all" names. Thus, we have that a does not co-refer with itself. But this is absurd! Hence, we have proven the negation of what we stipulated. We have proven that, for every name x there is a name y that actually refers to an object and co-refers with x. This is how we read the elements of our formalism. In intuitive sense, this means: every referent we attempt to name in our domain, exists.

Should this be a logical truth? Why is it a matter of logic what exists – or that "everything exists?" Is there anything that can be changed about our formalism, so that the above proof fails?

I.8 LOGICAL TRUTH, MODELS AND COUNTERMODELS, DECIDABILITY AND THE LÖWENHEIM RESULT

When we extend Propositional Logic to construct predicate or first-order logic, we make available symbolic resources that allow us to symbolize and prove validity for certain argument forms which fail to pass the validity test for Propositional Logic. This is not paradoxical: the paucity of expressive resources is responsible for the failure; there is nothing wrong with the propositional formalism and its restrictions are well-understood once we take into account what resources it has available to it. We pay a price, however, or so it may be argued, when we come to encompass more valid argument forms by ascending to the realm of predicate logic: we have now available to us resources that permit us to take stock of *non-logical* entities, like predicates and terms. It is not easy to define what counts as a logical notion – perhaps not surprisingly for the student of philosophic problems. One idea, due to Alfred Tarski and inspired by the study of geometrical theories, is that logical notions are precisely those that remain invariant under any possible transformations. We encounter a pleasing instance of this when we study functional completeness in the context of Propositional Logic. Interdefinabilities connecting the logical connectives of Propositional Logic permit us to choose building our system by using different sets of functions symbols which, if they are functionally complete, characterize exactly Propositional Logic: regardless of which such set we choose, we are guaranteed to have exactly the same logical system – or a notational variant of the same logic. Matters become more complicated in the context of predicate logic. We could even threaten mischief by pairing definitions of novel connectives to the kinds of contextual information that predicate logic can model. For instance, we could define a connective, which we can name "night owl", as conjunction if and only if there are predicate symbols for night owls in a predicate model's signature and defined as exclusive disjunction if and only if there are no predicate symbols for night owls in the model's signature. Is this connective a logical notion? It is not invariable under transformations, apparently.

This is an advanced and evolving subject and we will not pursue it in the present text but we can turn our attention to what happens, for instance, when we deal with predicate symbols – which, tellingly, are sometimes called *non-logical* constant symbols. These do not seem to be logical notions. Except for the distinguished binary predicate of identity, we cannot discover logical truths about the properties which are interpreted through predicate symbols. Obviously, in our predicate models we are able to construct narratives that have a decidedly empirical flavor about them. Of course, these are more like science-fiction narratives, not investigative forays into the empirical realm, but the point remains that we have strayed from the pure world of Propositional Logic in which modeling trades exclusively in referents that are logical meanings only by being truth values. The predicate symbol, on the other hand, relies on the abstract machinery of sets – which generates interesting issues of its own – but, more importantly, has as referent or assigned logical meaning a set (the extension of the predicate) which is open-ended and can be variable across models – ranging from the empty set to an infinite set. Yet, we can argue that we have not abandoned the territory of logical formalism in spite of this suddenly embarrassing availability of flexible narratives.

The key to understand why we are still squarely within the castle of logical analysis and formal construction is the notion of logical truth. In predicate logic – when undertaken by reliance on modeling – *truth is relativized to a model.* Notice, however, that we are not saying that *logical* truth is relativized to a model. This is what makes the whole difference. That truth value assignment is model-relative should be obvious by now. We can construct one model in which the proposition expressed by the statement "at least one student is industrious" is true and we can construct another model in which this proposition is false. There are propositions, however, that are true – and others that are false – across every logically possible model we can construct; or, we can say, they are true, or false, regardless of the relativization of truth to modeling. These are the logical truths (or tautologies) and the logical falsehoods (or contradictions) of predicate logic. Because there is a widespread view that predicate logic is required at a minimum to express the language of Mathematics, this is a matter of considerable gravity: it would be troublesome if we could not count on there being logical truths, and logical falsehoods, in predicate logic. But we can reassure ourselves on this score. A logical truth is a proposition that is true across all, or regardless of, logically possible models; and a logical falsehood is a proposition that is false across all, or regardless of, logically possible models. Given the neat partition of truth values into true and false – with negation defined as switching true to false and false to true – we can still count, of course, on logical falsehoods being the negations of logical truths and vice versa. Logically contingent propositions, on the other hand, (also called logically indeterminate or logically indefinite) are the propositions that can be possibly true and possibly false: we don't need to agonize over the loaded meaning of "possibly" or "logically possibly;" we have an immediate way out in that constructible models present – and are considered to exhaust – all logical possibilities. Thus, a proposition that can have models constructed for which it is true and other models for which it is false is a proposition that is possibly true and possibly false – and this is, by definition, a logically contingent proposition. No proposition can be both true and false in the same model – the semantics of our formalism forestall such an option. Thus, "possibly true and possibly false" is not a contradiction – as it should not be – since it simply tells us that our contingent proposition is true in some and true in some *other* models.

Aristotle, as the first thinker to engage in formal study of logic, had the habit of stating logical truths in predicate form. (His first statement is usually about how nature somehow upholds or verifies the logical truth, a view that is not accepted anymore.) Thus, Aristotle's statement of the logical truth of non-contradiction would take the form "nothing can be both true and false of something at the same time and in the same respect." This is a paraphrase, but we recognize the insistence on deploying analyzed propositions in which predicates are bonded with terms. The references to "the same time" are to be ignored, as we are dealing with so-called eternal propositions in the case of predicate logic – as we did for Propositional Logic: this means that all context that could have an effect on truth value is considered to be already telescoped into the propositional content. For instance, "John is a student" is understood to mean "John is a student at such-and-such-a-time or time-period, in such-and-such-collection-of-locations, and so on" for all variable contexts. To explicitly express dynamic contextual elements we require the symbolic resources of modal logic – something that is beyond our current ambitions in this text.

Let us take the following sequent and see if this a logical truth. We will consider how we can argue about this matter.

$$\models \forall x(Fx \vee \sim Fx)$$

To be a logical truth – thus, implied even by the empty set of assumptions as shown by the sequent above – this formula has to receive the truth value True in every constructible model of predicate logic. The easy way to proceed is by reductio: let us assume provisionally, for the sake of reductio, that this is not the case. Then, there has to be at least one model that is constructible, given our semantic arrangements and specifications, in which the regulating signature makes this formula False. Thus we have the following – and we move by using established rules for converting or interchanging quantifier symbols and also by using such standardly valid laws, we know from Propositional Logic, as DeMorgan's and Double Negation Elimination. Let us call our falsifying model ꟊ and its valuation or signature $\surd$.

$$ꟊ, \surd \models \sim \forall x(Fx \vee \sim Fx)$$
$$ꟊ, \surd \models \exists x \sim (Fx \vee \sim Fx)$$
$$ꟊ, \surd \models \exists x\ (\sim Fx \cdot \sim \sim Fx)$$
$$ꟊ, \surd \models \exists x\ (\sim Fx \cdot Fx)$$

This, however, is precluded by our semantic arrangements. For our presumed model to validate the existential formula, it would have to be the case that at least one constant denotes a domain object that both belongs and does not belong to the extension specified by the signature of the model for the predicate ⌜ F ⌝.

$$\text{For some } \lambda \in D(\mu), \surd(\lambda) \in \surd(F) \text{ and } \surd(\lambda) \notin \surd(F)$$

This says that, for some constant symbol, the referent of the constant in the model's domain (its value according to the model's signature), both does and does not belong to the extension (the value, according to the model's signature) of the predicate symbol. In this informal proof, we can piggy bag on the fact that our set-theoretical apparatus itself is strictly binary, the way it is set up, and it is nonsensical to speak of any abstract object as both being and not being a member of a specified set. Thus, we have reached logical absurdity and, accordingly, we must reject the assumption we laid down for reductio. In this way we have:

$$\mu, \surd \nvDash \sim \forall x(Fx \vee \sim Fx)$$
$$\mu, \surd \vDash \sim \sim \forall x(Fx \vee \sim Fx)$$
$$\mu, \surd \vDash \forall x(Fx \vee \sim Fx)$$

In other words, any arbitrary model we take cannot possibly falsify the given formula. The formula must, then, be true in every constructible (logically possible) model and, as such, it is, by definition, a logical truth – not simply true relative to some model or other but true as such or true across all, or regardless of any, specified model. In this way, we are satisfied that relativization of truth to models has not impacted our seminal ability to define logical truth. Hence, we are still dealing with proper logical formalism when we ascend to predicate logic.

It is important to be able to falsify given formulas – insofar as they are not logical truths. Or, given argument forms, it is crucial that we can construct counterexamples that, under the assumption that all the premises are true, falsify the conclusion of the argument form. Our recourse to construction of predicate models affords us an opportunity to do this. Any introduction to formal logic would have to begin by defining validity of argument forms. This is the holy grail of deductive reasoning. Definitions may speak of the truth of the premises "guaranteeing" that the conclusion is true – but this is rather imprecise talk, drawing on a vague and loose term like "guaranteeing." We need terms like "logically necessary" and "logically possible", which propel us into additional difficulties as these terms are not primitive; we risk circularity – if we use our definition of validity to define logical necessity – and we also risk casting forbidden glimpses in logics that actually extend propositional and predicate logic – the so-called modal logics that tackle systematically and formally the notions of necessity and possibility. Such notions can be assimilated to intuitions by using some appropriate modeling machinery: so, in Propositional Logic, the row of a truth table can be considered as a logically possible state of affairs and, in the case of predicate logic, as we saw, models with their signatures over specified domains play the same role of making available views of logical possibility. The constructability of at least one – thus, any one – model that invalidates a given argument form suffices to furnish a counterexample to this argument form. Of course, any argument form that can have at least one counterexample is invalid by definition – and the definition of invalidity in terms of availability of counterexamples may be said to have certain advantages. When we are able to construct a model that falsifies a presented putative argument form, we say that we have constructed a countermodel to the given argument form. Accordingly, a countermodel to a given putative tautology shows that the given formula fails to have the logical status of tautology.

Models can also be constructed to establish consistency or joint satisfiability: this means that some model should be constructed in which all formulas of a given set of formulas receive the truth value True in accordance with the model's signature. This suffices to establish consistency – it shows that all formulas can possibly be "true together" or they can be true in a model. Characteristically, this model is also called sometimes a "theory" of the given set of formulas. If the given set can have a theory – which means, a consistent theory – then the formulas constitute a consistent set.

To establish logical status of falsehood – that is, logical falsehood and not simply falsehood in some model – we can take the negation of the given formula and establish that it cannot be falsified – it cannot have any countermodel. We know that tautologies and contradictions are mutual contradictories – this remains the case in predicate logic as it depends on the availability of exactly two truth values one of which only satisfies and also on the definitions of the connectives, which remain the same in predicate logic.

We can also use modeling to establish, or refute the claim, that a given set of predicate formulas is consistent. Construction of at least one model in which all the formulas receive the valuation True shows consistency. Construction of at least one model in which not all the formulas are True constitutes a countermodel to the consistency claim – and, thus, it serves as a model for showing that the given formulas are inconsistent when taken together as a set.

As an example, we are given the following formulas which express Irreflexivity and Symmetry. A binary relation is irreflexive if and only if each member of the domain is not R-related to itself. A binary relation is symmetric if and only if, for every pair of objects in the domain, if it is true that the first is R-related to the second, then it must be true also that the second is R-related to the first.

$\vDash ?\ \{\forall x \sim Rxx,\ \exists xRxx \cdot \exists y \sim Ryy\}$

$\mathcal{M} = $ <D = {①, ②},
√ = {{<a, ①>, <b, ②}, {<①, ②>, <②, ①>}}>

Thus, we have:
It is not the case that <①, ①> or <②, ②> is in the valuation or extension of R:
Therefore, it is true that,

$\sim \exists xRxx \equiv \forall x \sim Rxx$ (Irreflexivity)

But both <①, ②> and <②, ①> are in the extension of R. Thus:

$(Rab \supset Rba) \cdot (Rba \supset Rab)$

Therefore,

$\forall x \forall y(Rxy \supset Ryx)$ (Symmetry)

Having constructed a model in which both given formulas are true, it follows that the set of these two formulas is consistent. Standing back from the formal language, we can then say that Irreflexive and Symmetry are consistent mutually, as properties of relations. A relation can possibly be both irreflexive and symmetric. (We would need the symbolic resources of second-order logic, which allows to quantify over predicate letters and also gives symbols for predicates of predicates, to express this formally.)

Finally, we can use modeling to establish independence of a formula relative to one or more other formulas. A formula φ is defined as independent of formulas $\psi_1, \psi_2, \ldots, \psi_n$ if and only if it is logically possible that (i.e. there is at least one constructible model in which) all of $\{\psi_1, \psi_2, \ldots, \psi_n\}$ are true and φ is false.

As an example, we will construct a model that establishes the mutual independence of the properties of properties, which are known as Weak Connectivity and Transitivity. To show mutual independence of two formulas, we have to show that each one of the two formulas is independent of the other – in the sense of independence we defined in the preceding paragraph. This might need two different models.

(Strong Connectedness) $\forall x \forall y((x \neq y) \supset (Rxy \vee Ryx))$
(Transitivity) $\forall x \forall y \forall z((Rxy \cdot Ryz) \supset Rxz)$

𝓜 = <D = {①, ②}, √ = {{<a, ①>, <b, ②>}, {<①, ①>}}>
This model falsifies strong connectedness insofar as neither one of the objects in the domain are R-related to each other. On the other hand, it verifies transitivity, which requires some effort to discern. Since only one object is related to itself, the conjunction that makes the antecedent of the transitivity formula is true:

$\surd(Raa \cdot Raa) = T$

The truth value of the consequent is also True.

$\surd(Raa) = T$

In all other cases, which we can verify by developing the formulas as we have learned how to do, we have false antecedents and thus, for the standard definition of the implicative connective, we have vacuously true implication formulas. For example:

$\surd((Raa \cdot Rab) \supset Rab)$
$\surd((Rab \cdot Rba) \supset Raa)$
Etc..

This model establishes that transitivity is independent of strong connectedness.

We can also construct some model – not the same as the preceding one – to show the other side of the independence claim: this model will have to confirm strong connectedness and falsify transitivity, in this way establishing that strong connectedness is logically independent of transitivity. Here is such a model.

𝓜 = <D = {①, ②}, √ = {{<a, ①>, <b, ②>},
{<①, ②>, <②, ①>}}>

In this model we have:

√((Rab ∨ Rba) = T since < √(①), √(②)> ∈ √(R) and
<√(②), √(①)> ∈ √(R)

On the other hand, transitivity fails, for instance, in the following case:

√((Rab · Rba) ⊃ Rbb) = F

The antecedent is True, as established above, but the consequent is False since it is <u>*not*</u> the case that <√(②), √(②)> ∈ √(R)

Two logical models 𝓜1 and 𝓜2 are logically equivalent,

𝓜1 ≈𝓜2

If and only if they verify exactly the same formulas (which means that exactly the same formulas are true in the two models):

Iff ∀ϕ(𝓜1⊨ ϕ ⇔ 𝓜2 ⊨ ϕ)

Brief reflection should show that we can assign different interpretative keys for two equivalent models. Conversely, if different interpretative keys leave invariable the truth values of all the formulas of two models, then the two models are logically equivalent. This shows us, again, that the assignments of interpretations for semantic purposes are not themselves logical notions since they do not remain invariant along with the other elements of an altered model. Interestingly, if we come across any non-logical expressions that do not show the same behavior – which means that they cannot be varied without altering model equivalence – we have stumbled on notions that are not expressible with the resources of first-order predicate logic.

As example, we start with a model ꟊ = <D, √> as follows:

$\mu = <D = \{o_a, o_b, o_c\}, \surd = \{\{<a, o_a>, <b, o_b>, <c, o_c>\}, \{F, R\}\}>$

$\surd(F) = \{o_a\}$

$\surd(R) = \{<o_a, o_a>, <o_b, o_b>\}$

KEY1: Fx: x is a student; Rxx: x likes himself.
KEY2: Fx: x is a teacher; Rxx: x dislikes himself.
In spite of the drastic change of subject matter in the interpretative narratives assigned to the model, the model essentially remains the same.
Now consider the following model which is equivalent to the one presented above.

$\mathcal{M} = <\{1\}, \surd = \{\{<a, 1>\}, \{G, S\}\}>$

$\surd(G) = \{1\}$

$\surd(S) = \{<1, 1>\}$

The two models μ and $\mathcal{M}$ are logically equivalent.

We conclude by briefly introducing a seminal result obtained by the German logician Leopold Löwenheim in 1925. It is significant that this result, which ensures decidability in a semantic sense we will specify, applies only in the case of monadic predicate logic and is forfeited when identity and relational (higher than unary) predicate symbols are added. According to the Löwenheim result: *if a countermodel exists for a given well-formed formula of monadic predicate logic, then it is guaranteed to be generated by means of a model that has at most 2^n elements, where n is the number of distinct predicate letters in the given formula.*

I.9 EXERCISES

(1) We can evaluate whether a given set of predicate formulas is consistent by constructing at least one model in which all given formulas are true. In the semantic expression of Propositional Logic, modeling is undertaken by means of truth tables – also by using the tree method we studied – and the availability of a satisfying model is a row of the truth table for given formulas, across which, row, all the

formulas receive the valuation true. This shows that they express propositions that can all be true together. If the truth table for the given formulas has no row across which all the formulas are true, this establishes inconsistency or unsatisfiability – since the truth table is provably an effective decision procedure we can use reliably in Propositional Logic. For the tree method, at least one open, non-closed, path settles truth values for the atomic letters, for which all the given formulas are true: hence, this establishes consistency. If all paths are closed upon termination of the tree (which, provably, terminates after a finite number of steps in rule application), then the determination is that the given set of formulas is inconsistent. In the case of Predicate Logic model-construction, we can show consistency by constructing at least one model in which all the given formulas are true. Failure to achieve this, however, does not establish that no such model is constructible.

A stricter definition of inconsistency has it that a set of propositional formulas is inconsistent if and only if a contradiction – or any propositional formula whatsoever – can be proven validly from the set.

Can you construct <u>satisfying/verifying models</u> for the following sets of formulas – showing in this way that the given set is consistent or jointly satisfiable?

a. $\{\forall xRxx, \exists xRxx, Fa, \exists xRxa\}$
b. $\{\exists x\exists y(x \neq y \supset (Fx \cdot Fy)), \forall x(Fx \supset Rxx)\}$
c. $\{f(a) = b, \exists xFx, \exists xFf(x)\}$
d. $\{\forall x(Fx \supset Gx), Fb, Gc, \sim \forall xGx\}$
e. $\{\forall x(Fx \supset Gx), \sim Ff(a), \exists xRxf(x), \sim\exists xRxf(x)\}$
f. $\{Rab, \exists x\exists y \sim Rxy, \forall x\forall y(Rxy \supset Ryx)\}$
g. $\{\sim \exists xRxx, \exists x\exists y(Rxx \equiv Rxy\}$
h. $\{Ff(a) \supset \exists xRf(x)a\}$

i. {∃x∀yRxy ≡ ∀x∃yRxy}
j. {Raa ⊃ ∃x∃y(x = y · Rxy)}
k. {Rf(a)f(b), ~ Rab, ∃x∃yRf(x)y}

(2) Examine if the given models verify or falsify the appended formulas.

a. μ = <D = {△, □, ◊}, √ = {<t, △>, <s, □>, <d, ◊>,
<F = {□, △, ◊}>>
⊨? ∀x∀y∀z(Fx ≡ (Fy ≡ Fz))
b. μ = <D = {↺, ↻},
√ = {<l, ↻>, <m, ↺>, <R = {<↻, ↺>, <↻, ↺>}}>
⊨? (Rll ≡ Rmm) ≡ ∀x∀y(Rxx ≡ (Rxy ≡ Ryx))
c. μ = <D = {❽, ❿},
√ = {<a, ❽>, <b, ❿>, <f(❽), ❽>, <R = {<❽, ❽>}>}>
⊨? ∃x(((Rxx ⊃ Rf(x)f(x)) ⊃ ∃y(x ≠ y · ~ Rf(x)y))
d. μ = <D = {①, ②, ⓫, ⓬},
√ = {<a, ①>, <b, ②>, <c, ⓫>, <d, ⓬>,
<f(①), ⓫>, <f(②), ⓬>},
<F = {f(②), ①}>,
<R = {<⓫, ⓬>, <①, ②>}>}>
⊨? ∀x∃y(Ff(x) ⊃ Rf(y)x)
e. μ = <D = {⋈, ⋉, ⋊}, √ = {<k, ⋈>, <l, ⋊>, <m, ⋉>, <n, ⋈>, <F, {⋈}>, <G, {⋉, ⋊}>}}>
⊨? ∃u∃v(u = v) · ∀x(Fx ≡ ~ Gx)
f. μ = <D = {⊶, ⊷}, √ = {<a, ⊶>, <b, ⊷>,
<R, {<⊷, ⊷>}}>
⊨? ∀xRxx ∨ ~ ∃x∀y(Rxx ∨ Rxy)

(3) Construct, if possible, countermodels to the shown argument forms of predicate logic. Taking logical truths (tautologies) to be no-premises conclusions of valid argument forms, we can also ask for construction, if possible, of countermodels to claims about tautologies

of predicate logic. Such purported formulas – to be examined for tautologousness status – are represented as conclusions of argument forms with the empty set of premises.

Make use of the Löwenheim Result to decide what cardinality degree you need to ascend to for your models' domains. (Or, we could say, use the result to determine your maximal model cardinality because it is also customary to speak of a model's cardinality which is determined, naturally, as identical with its domain's cardinality.)

a. $\exists x \exists y Rxy \models \exists x \exists y (x \neq y \cdot Rxy)$
b. $\exists x \exists y (x \neq y \cdot Rxy) \models \exists x Rxx$
c. $\forall x \forall y Rxy \models \sim \exists x Rxx$
d. $\exists x Rxb \models \exists y Rby$
e. $\emptyset \models \exists x \exists y Rxy \equiv \forall x \forall y (Rxy \equiv x \neq y)$
f. $Rf(a)f(b) \models \forall x \forall y Rf(x)f(y)$
g. $\emptyset \models \exists x (Fx \supset \exists y (y \neq f(y) \cdot Rxy))$
h. $\forall x Rxa \supset (\sim Raa \vee \exists y Rya) \models Rba$
i. $\exists x Rxf(x) \models \forall x \forall y (x \neq y \supset Rxf(x))$
j. $\forall y Fy \models \exists x \exists y (x \neq y \cdot (Fx \equiv \sim Fy))$
k. $a = b \vee f(a) = f(b) \models \forall x \exists y (Rf(x)y \vee Rf(y)x)$
l. $\emptyset \models \exists x (\exists y Rxy \supset \forall z Rxz)$

II. TRANSLATIONS FROM NATURAL LANGUAGE INTO THE FORMAL LANGUAGE PRED

Translations from English, the target natural language, will be undertaken by using the expressive symbolic resources of PROP, the formal language whose grammar we have already given. We add, however, capital letters, possibly with subscripts from the positive integers, as we did with the grammatical setup of L, in order to translate propositions expressed by declarative sentences of English. (Contrast the use of the small letters, with or without subscripts, for symbolizing propositions as parts of logical forms.) Now, an interesting issue arises. Obviously, although PRED retains individual propositional variables, such expressive means will not be needed, for the most part, since we have access to symbols for representing the inner parts of propositional contents. The question may be asked: should we not stipulate separate symbolic resources for translating internal parts of English sentences – as distinguished from the symbols we use for logical forms? Strictly speaking, the answer is affirmative since different abstract items are rendered by the various symbolizations. Such a move is almost never encountered in textbooks. The cause of this omission – which we will respect even if reluctantly – has to do with convenience. Ambiguity should never arise and we expect to accomplish this by making the appropriate disclaiming statements at this point. We will say that the symbols happen to be the same – as a matter of coincidence, so to speak – but the things that are symbolized are different. Another issue is that, rarely, we could run into a situation in which a propositional variable

– a capital letter – is the same as one for a predicate constant (if we allow, for instance, for a key of translation that takes the initial letter of the linguistic part to serve as symbol and this symbol happens to be the same with one chosen for a propositional variable. The disambiguation we need is readily available, however, in that only the predicate symbol, unlike the propositional variable, is accompanied by individual variable or variables letters. Superscripts indicating arity of predicate and function symbols may be omitted and, as usual, parentheses are to be used only for disambiguation.

GRAMMAR(PL): A_j/A_k, ..., F_i,G_l,.../x_m, .../a_j, .../{~, ·, ∨, ⊃, ≡}/{∀, ∃}/{=}/f_j,...//(/)

This is one of the most intricate tasks we can undertake in the context of studying the standard logic. Delivering on the pledge that the formal systems we develop can be put to use in examining the logical structure of linguistic constructs, we must specify how to translate from natural language – English in our case – into a formal idiom like PRED. We assume knowledge of how to translate into the propositional idiom PROP, which we have extended for generation of PRED. The inability of translating non-truth-functional parts of sentences is also marked and it applies when we translate into predicate logic idioms as well. Non-truth-functional expressions generate what are called opaque contexts: this notion means, by definition, that logically equivalent propositions are not inter-substitutable into such contexts in such a way that the truth-value is preserved (salva veritatis.) Even though both propositions expressed by the sentences "It is often sunny in Arizona" and "triangles have three angles" may be accepted as true, substitution into the context generated by "it is necessarily the case that__" will yield true only for for the logically-necessary proposition about triangles. This serves as an example of a referentially opaque context. The word "referentially" is to be understood as significant here because the semantics of extensional logic requires, foundationally, that logical meaning is assessed as a matter of an ascertainable reference to a specifiable entity (without regard for its metaphysical status), which is called the referent or denotatum. Propositions have truth values as referents, non-logical constants like refer to sets of entities, logical terms may refer to entities in a domain, and so on. Philosophically, this has repercussions because it banishes claims that cannot be evaluated in terms of a specifiable referent. Intensional logic, which falls outside of our present purview, extends into regions of formal studies that can manage referentially opaque contexts but even in that case certain extensionalizing devices are used. The interested reader should undertake the study of modal logics in this regard. To sum up, predicate logic translations have no remedy when it comes to capturing phrases that generate referentially opaque contexts. We cannot use predicate logic resources for this purpose, which also tells us that Predicate Logic remains solidly within the ambit of extensional formalisms. Indeed, any weakening of the principle of inter-substitutivity of logically equivalent propositions would mean, automatically, that we are not dealing with an extensional logic; this is not the case with predicate logic.

The predicate logic system PRED extends PROP, the Propositional Logic language, and, for this reason, inherits the same logical connectives. If we restrict our observations to finitary domains, we can consider the added symbolic resources for the quantifiers as representing additional connective symbols. We do, however, intend to effect translations, if need be, for infinite domains – like those that are evoked by the language of Mathematics. The comments and instructions given in 0.4.4 regarding translations into PROP are presumed to be carried over integrally into the present context. We cannot bypass the problems we investigated, stemming from the truth-functional character of the connectives of the logical system. Of course, we have enhanced symbolic resources and are able now, in PROP, to translate inner parts of meaningful sentences of a language like English. We should think of our translating powers as being reinforced vis a vis the capabilities of PROP. By the same token, we inherit now problems that are specific to translational limitations that can become visible only when resources are available for rendering formally such parts of natural language as pronouns, names, verbs, quantifier phrases, and so on.

Certain preliminary observations, which are quite standard, have to do with how we translate certain expressions of the natural language. We only have two quantifier symbols: the universal and an existential quantifier – interdefinable with the universal quantifier – which means "there is at least one such-and-such." We will see later that this can create certain problems. Clearly, we do not have (what would be non-standard) quantifier symbols for such expressions as, for instance, "a few are F" or "most are F."

A natural language like English does not accommodate our formalist aspirations by compelling that phrases wear their logical conditions on their sleeves - so to speak. The proposition expressed by such a sentence as "the soldier faces dangers" is one that requires universal quantification and is to be understood as "for all x, if x is a soldier, then x faces dangers." This is the same translation we would need to enforce for "all soldiers face dangers," "any soldier faces dangers," "a soldier faces dangers," and "anyone who is a soldier faces dangers." A staple of textbook instructions at this point is to draw attention to the fact that the logical grammar of universally quantified propositions - not obvious in the surface grammar of sentences of a language like English – compels us to use the implication connective symbol: "all F are G," has as its logical grammar "for all x, if x is an F, then x is a G."

On the other hand, it must be emphasized that existentially quantified statements must be rendered – generally – by means of the conjunction connective symbol. Failure to do so, as we will see, inflicts distortion of meaning. Thus, "some F are G" is to be understood in its logical grammar as "there is at least an x such that x is F and x is G." If we use the implication connective, taking this to mean "there is at least an x such that if it is F then it is G" we run into the prospect of having the expressed proposition as being vacuously true in case it is false that "there is at least an x such that x is F." Clearly, however, the competent speaker's intention is to take the proposition as false in the case in which it is false that some x is F.

Of course, a proposition like the one expressed by the sentence "if something is red, then something is colored" will be rendered as an implicational proposition: but, notice that in this case the main connective is itself the implication connectives whereas in the preceding example the main connective (if we are speaking of finite domains) is the existential quantifier symbol itself.

II.1 TRUTH CONDITIONS AND DEEP STRUCTURE

We can consider a translation to be successful only if the target proposition, which is translated into our formal language, has the same truth conditions with the formula that translates it. For sets of propositions, the truth conditions under which they are all true together should match exactly the truth conditions under which the set of translating formulas is consistent or satisfied. For arguments, validity should be preserved under translation in the sense that the argument form instantiated by the formal rendering of the argument is valid or invalid if and only if the translated argument of the language is, respectively, valid or invalid.

Between any two translations that have the same truth conditions, and render the same proposition, the one that shows a relatively deeper logical structure should be considered as more successful. For instance, while we can translate by means of a unary predicate constant interpreted as "x gives something to someone" and symbolized, for instance, by "G^1x," we exhibit a deeper logical structure when we translate, instead, by means of a binary predicate, symbolized by "G^2xy" and interpreted as "x gives something to y." Clearly, by the same reasoning, and given the implied criteria for assessing structural depth, choice of a ternary predicate constant symbol, "G^3xyz ", interpreted as "x gives y to z", would take us to even deeper logical structure. Logical structure, then, is a matter of showing as much detail as possible – with respect to what can be captured by the symbolic resources of a formal language, of course.

There may be cases, however, in which striving to show a deeper structure is defeated, and cannot be done without running into translational failure, because of certain limitations we face. A classic example is the phrase "small elephant" which should not be rendered by using two predicate symbols, one for "small" and one for "elephant." Consider that a small elephant might not be small as such; if, however, we symbolize conjunctively, we would be able to derive validly the proposition stating that the entity we are talking about is small if it is given that it is "small and an elephant." On the other hand, and underscoring the characteristic type of difficulty we run into in predicate logic translation, we do want to derive validly from the proposition about something being "a small elephant" that it is "an elephant" but it appears that the criticism made above prevents us from symbolizing in such a way that we are able to make this derivation.

The requirement that the deepest possible logical structure should be shown in translation should be understood as constrained by whatever overriding limitations we face. Moreover, domain restrictions, or absence of restrictions, play a role as well. If we choose an unrestricted domain, we have to render "someone gave something to someone" by "there is someone, x, a person, and there is something, y, a thing, and there is someone, z, a person, and x gave y to z." Thus, it seems that unrestricted domain stipulations compel us to represent a relatively deeper structure of the target proposition. This does not mean, however, that we should always elect a restricted domain. It depends on what other considerations come into play.

We give examples of preferential translations relative to the consideration about showing deeper structure.

1. (S1) Everyone who is a student and is at the door will be admitted.

   a. Key: Dx: x is a student at the door.

   Ax: x is admitted.

   $\forall x(Dx \supset Ax)$

   b. Deeper Structure: Key: Sx: x is a student;

   Dx: x is at the door;

   Ax: x is admitted.

   $\forall x((Sx \cdot Dx) \supset Ax)$

2. (S2) Anyone who is a father has some child.

This case offers us the options of using predicate symbol or function symbol to render "x is the father of y." We may wonder if it makes any difference from the standpoint of structural depth that is captured by the translation. But there are some other possibilities too that allow us to consider which one of our relative options we should adopt.

   a. Key: Fx: x is a father.

   Cx: x is a child.

$\forall x(Fx \supset \exists yCy) \equiv \forall x\exists y(Fx \supset Cy)$

b. Deeper Structure: Fxy: x is a father of y.

Cxy: x is the child of y.

$\forall x(\exists yFxy \supset \exists zCzx)$

We might observe that the virtue of greater simplicity of the preceding translation may be preferred in spite of its not showing as deep a structure as the second translation; the consideration we need to attend to is: what is lost and how important is that? If we choose the less deep translation, we cannot derive validly that there is someone who is the child of the person who is father.

c. Key: $f(x)$ = the father of x.

$c(x)$ = the child of x.

$\forall x(\exists y(x = f(y)) \supset \exists z(z = c(x))$

If we are wondering as to which one we should prefer between b and c: for most purposes, it should not make any difference.

We are able to symbolize temporal references – insofar as the temporal phrases do not generate referentially opaque contexts in the specific linguistic usage we are targeting for translation. For example, the proposition expressed by the sentence "there will be a sea battle tomorrow" has a non-truth-functional phrase in "there will be__" which we could not render by means of any truth-functional connective in Propositional Logic. None of the four mathematically available unary truth-functional connectives can be used. Given the resources of predicate logic, we still have only truth-functional connectives at our disposal. We can, however, remove restrictions on the domain and construct predicate symbols for moments and temporal relations. We can even opt for detail, risking cumbersome expressivity, as in the following translation.

(S1) There will be a sea battle tomorrow.

There is a moment x and there is a moment y and x is tomorrow and y is tomorrow and x is before y and a sea battle starts at x and ends at y.

$\exists x \exists y(Mx \cdot My \cdot Tx \cdot Ty \cdot Bxy \cdot \exists z(Bz \cdot Szx \cdot Ezy))$

The predicate symbols that share "B" are disambiguated because one is unary and the other binary.

We can give more examples in which temporal references are rendered symbolically.

(S2) It never rains in the plain in Spain.

KeyL Tx: x is time period;

Pxy: x is a plain in country y;

Rxy: it rains at place x at time period y;

s: Spain.

$\sim \exists x(Tx \cdot \exists y(Pys \cdot Ryx))$

(S2) Only when it rains do the flowers thrive.

Key: Fx: x is a flower;

Tx: x is a time period;

Vxy: x thrives over period y;

Rx: it rains at x (x is rainy).

As we know from the translation of Propositional Logic, "only" introduces the necessary condition and is, in translation, placed after the conditional or implicative symbol. We can also show logically equivalent formulas and we can translate back into English. The golden rule is to opt for the translation that is closer to the structural arrangements of logical particles in the sentences we are translating (which is the first formula in the left.)

$$\forall x \forall y((Fx \cdot Ty \cdot Vxy) \supset Ry) \equiv \forall x \forall y(Fx \supset (Ty \supset (Vxy \supset Ry))) \equiv$$
$$\sim \exists x \sim \exists y(Fx \cdot Ty \cdot Vxy \cdot \sim Ry) \equiv \ldots$$

II.2 DOMAIN SPECIFICATIONS AND CAPTURING CONTEXT

The propositions of the standard logic are sometimes called "eternal propositions," a term that could, wrongly, conjure up far-reaching metaphysical connotations. What this means is that the proposition is considered to be including descriptions of all dynamic elements of context that could affect the truth value to which the proposition refers: this ensures that contextual elements – like indexicals, for instance, or the flow of time – cannot play a role in determining the proposition's meaning (truth value.) Instead of dealing with a proposition like "Socrates is sitting" which may be considered as true at all times at which Socrates may be said to be sitting, and false otherwise, we have in our system symbols for a proposition understood to consist in "Socrates sits at such-and-such a time and in such-and-such a location, and..." – and so on for all the context-sensitive or dynamic elements. This applies in the case of predicate logic as well.

We may wonder if we can use the resources of predicate logic to allow ourselves to present some contextual elements, like temporal instants, for instance. This would not affect the restriction of our propositional types – as "eternal propositions" – but it would allow us to render certain expressions like "always" and "sometimes." The domain would have to be made unrestricted so that predicate letters can be assigned for temporal moments and for persons and whatever else we have in the target sentence. Other contextual elements can also be treated in this fashion. Still, the remaining contextual elements – an open-ended variety of them, indeed – are considered packed into the proposition - so we continue to deal with what we called "eternal propositions."

We can now give examples.

1. (S1) John is distracted sometimes.

a. We can say that this translation has shallower structural depth – in reference to our preceding discussion. The domain is unrestricted so that entities can be characterized by use of interpreted predicate letters. While the person referred to by "John" is a person, in our example, we also need to characterize entities that are time-periods.

Key: Dx: x is distracted sometimes.

j: John.

Dj

b. Greater depth: Key:

Tx: is a time-period.

Dxy: x is distracted during y.

j: John.

$\exists x(Tx \cdot Djx)$

2. (S2) You can never deceive educated people.

We cannot render the modal phrase "it is possible, in some sense, that__"; so "can" is non-truth-functional and eludes our translational capabilities. We take the meaning of the given sentence to be the same as the meaning of: "educated people are never deceivable."

a. Key: Dx: x is never deceivable.

Ex: x is educated.

$\forall x(Ex \supset Dx)$

b. Greater depth: Key: Tx: x is a temporal moment.

Ex: x is educated.

Dxy: x is deceivable at moment y.

$\forall x \forall y((Tx \cdot Ey) \supset \sim Dyx) \equiv \forall x \forall y(Tx \supset (Ey \supset \sim Dyx) \equiv \forall x \sim \exists y(Ex \cdot Ty \cdot Dxy) \equiv \sim \exists x \sim \exists y(Tx \cdot Ey \cdot Dyx)$

II.3 COMPACTING PREDICATES

We cannot express, for instance, adverbial modifiers by special symbolic means. We need to interpret predicate letters in such a way that the adverbial, or any other type of modifier, is compacted into the logical predicate. For instance, "John walks fast" has to be translated by means of a key that contains a predicate letter, for instance, "W" such that "Wx" is interpreted as "x walks fast."

We can generate compact predicates in the manner indicated in the preceding example but we are mindful not to forfeit a translation that shows lesser depth – something we addressed in the previous section. On the other hand, as with adverbial modifiers, we do not have any alternative options. To stay with the same example, "John walks fast" cannot be translated faithfully as:

Wj · Fj

The key is: Wx: x walks; Fx: x is fast.

This translation does not have the same truth conditions with the proposition of the given sentence because it is possible that John has the characteristics attaching to a person who is fast only when he walks; so, the conjunctive statement could be false when we aim at translating a proposition that is true. If we use as interpretative key, "Fx" means "x walks fast," then the proposition we translate is "John walks and John walks fast." Although rather unusual by textbook standards, this might not be a bad idea insofar as we are able to infer validly – as we should be able to infer – that "John walks." There is, however, a sense in which we could say that we have not translated exactly what we were given. It might be that John walks only when he walks fast; in that case, which cannot be excluded, we have mistranslated and, in so doing, we have, again, fallen into the trap of having a proposition that does not have the same truth conditions with the target proposition of the linguistic sentence we were given.

We have no symbols and no grammatical arrangements – and we have undertaken no matching semantical accommodations – for higher-order predicate symbols. This means, for practical purposes, that we cannot translate directly propositions like the one expressed by the sentence "some colors are brighter than other colors" with "brighter" being a property of properties and, as such, requiring second-order predicate letters – and quantifiers over our first-order predicate letters. We need, therefore, to squeeze this proposition into some formula that is available given our resources, keeping in mind that we might be unable to translate second-order tautologies by formulas that are first-order tautologies or to translate valid arguments so that we have, in our formal language, instances of valid predicate argument forms. It should be realized, however, that argument forms that are valid when translated into higher order predicate systems may be invalid in their first-order predicate translations. This is a general phenomenon we find as we cross over from Propositional Logic to predicate, from predicate to second-order, and also to modal logics. There are controversies surrounding non-standard logics but we cannot enter here into that subject. We should mention that truth-functional logics with more than two truth values represent alternative approaches to semantic concepts; although we do not discuss many-valued logics in this text, we can briefly refer to them as not necessarily making more resources available for translation but rather as changing the rules of the semantic game – so to speak – and affecting how we translate more radically. (For instance, we might become able to translate phrases like “John is sort of tall” or “Jill is quite tall” but we do this through a radical reconfiguration of the way we construct predicate extensions altogether. This is philosophically controversial, needless to say.)

To continue with our example, we wonder how we may translate the proposition of the sentence:

(S) Some colors are brighter than other colors.

Key: Cx: x is a color; Bxy: x is brighter than y.

$$\exists x(Cx \cdot \exists y(Cy \cdot Bxy))$$

Because it expresses an empirical claim, the proposition should be logically indeterminate – neither a tautology nor a contradiction. This is the case with the formula we use to translate the proposition. We cannot, however, validly infer from this proposition that there is a property such that some things have it so that they are brighter than other things which also have this property. The reason we cannot extract this intuitively valid inference is due to the limitations of our symbolic resources. We do not study second- or higher-order logics in this text. We should point out that there are some metalogical anomalies that accrue when we ascend to the investigation of higher-order logics.

Clearly, we also lack resources for translating modals, which would require an extension of the standard propositional or the standard predicate logic by adding modal operators. Usually, such operators are unary – although we do not have to confine ourselves to unary modals – and a successful semantical approach uses the notion of possible worlds that is due to Saul Kripke, based on an old idea that occurred brilliantly to the German philosopher Leibniz. Although, strictly speaking it lies beyond our present scope, we may point out that the metalinguistic justifications made available through the possible-worlds Leibniz-Kripke semantics is actually the language of predicate logic. (Its strength, however, nods rather in the direction of second-order but, interestingly enough its expressive power is only that of a fragment of the predicate logic – the fragment that remains invariant under a technique covered by advanced texts as bisimulation.) We wonder if we can compact modal claims by means of our first-order predicate logic resources. For the same reasons we mentioned above, we will not be able to translate so that tautologies and validity are preserved. As an example, consider the proposition made by the sentence:

(SModal) It is necessarily true that triangles have three angles.

The non-truth-functional modifier "necessarily" cannot be translated. There is a good question here as to whether we should translate with the full panoply of symbolic means made available in predicate logic – so that we have for an evident choice of interpretative key:

(Tr1) $\forall x(Tx \supset Ax)$

Or, should we rather bow to the non-truth-functional character of the modifying modal – whose scope ranges over the entire proposition we have translated above – and, accordingly, translate as:

(Tr2) A

The fact that the scope of the modal is all-encompassing appears to compel us to opt for the latter. Nevertheless, if the speaker's intent is to express the truth that all triangles have three angles, then this can be done by means of the universally quantified formula. This is an analytic truth – which means that the meanings of the words in the sentence compel valuation as True and regardless of any empirical considerations or verification. The words whose meanings play this decisive role in assigning the status of analyticity to the proposition are non-logical words (the words "triangle," "angle" and "three.") [Notice, incidentally, as a theoretical challenge, and try to explain why the copula or the verb "to be" used grammatically for joining is inert for purposes of our logical analysis.] If we translate by means of the universally quantified formula, we do not have a tautology but this is the case, in the first place, with the target proposition: this is an analytic proposition but only by virtue of the meanings of the non-logical words of the sentence that expresses it. So, we can argue that nothing is lost. Moreover, there is way to confer an honorary status of logical truth to the formula by using something – rarely covered in textbooks – that is called Meaning Stipulation. This is like legislating a stipulative definition within a theory. Indeed, in a theory like that of the Euclidean geometry, in which all proofs proceed from analytical truths to conclusions that are themselves analytical truths, definitions (along with axioms, definitions and already proven theorems) can be introduced as lines at any point in a valid proof. We treat the formula in Tr1 above as a meaning postulate if we legislate that it can be used as True at any point.

Case-by-case analysis is needed when we translate. In some cases, compacting is needed but in others it should not be undertaken. We give examples.

1. John teetered precariously over the edge of the precipice.

Here we need to take stock of depth-related considerations and opt for an unrestricted domain whose open-ended character compels us to generate predicate letters for "precipice", and so on. An interesting observation is that, if the precipice in question is distinguished in some way, depending on the context of the narrative, we should actually use an individual constant to name it.

Key: j: John.

Txyz: x teeters precariously over y at moment z.

Px: x is a precipice.

Mx: x is a moment in time.

Sx: x is in the past.

$\exists x(Px \cdot \exists y(My \cdot Py \cdot Tjxy)) \equiv \exists x\exists y(Px \cdot My \cdot Tjxy)$

2. Jack and Jill are delightful to have at a party.

The problem we face here is not about compacting although we might wonder if compacting of predicates may come to the rescue. This is an issue we faced when translating into the resources of Propositional Logic: we do not know if this is a conjunction indeed and, in many if not in most contexts, this is not a conjunctive proposition. The point may well be that Jack and Jill are delightful to have in a party together, and only when they are together, whereas each one of them might be morose and unsocial, for instance, if by themselves. In that case, we do not have a conjunction. Predicate compacting will not come to the rescue because the issue arises at the connective-level. A solution that might work would have us compact the subjects but we are unable to do this because we have no way of generating an individual constant symbol whose interpretation compacts more than one objects in the domain. (Actually, we can use identity to make two constants co-refer, as we know, but this does not solve our problem in this example.) Thus, the translation is:

A

Possibly, we might try this translation.

Key: Px: x is a party.

A: Jack and Jill are delightful to have.

$\forall x(Px \supset A)$

Let us conclude by noting that we cannot translate phrases that seem to point to penumbras or borderline areas – as when we have a sentence like "Jill is quite tall." Our predicate extensions, like the one for tall in some model or other, are "sharp" or "crisp", not "fuzzy": this means that for any member of the domain, either it is a member of the extension of the predicate or it is not. We do not permit for degrees of inclusion – as the case is for a controversial approach known as Fuzzy Logic. This means that we have to compact predicates – so that "quite tall" may be the interpretation of a predicate letter with "not so tall" being another. In this way, we risk translating intuitively valid arguments as invalid but we cannot undertake the drastic revision to standard logic that would be required in order to be able to precisify degrees of inclusion. Thus, "Jack is not tallish, therefore, he cannot be quite tall" cannot be symbolized by compacting predicate so that it is valid. We do pay this price as a matter of incurring the ineluctable formal limitations that any formal system impose.

II.4 NON-STANDARD QUANTIFIERS

We only have two quantifier symbols – one for the universal quantifier and one for a specialized existential quantifier which is to be understood as meaning "at least one." We can also think of the existential quantifier as definable by means of the universal: "not all not___," which gives us, rightly understood, "at least one" or "one or indefinitely more." In everyday language, we an abundance of encounter quantificational phrases which we cannot translate into our predicate idiom: a few, many, most, more rather than less, about the same as not, etc.... There are alternative logics that have formal means and semantical arrangements for accommodating the semantics of such expressions. We do not have such access in the standard predicate logic.

So, how should we translate the proposition expressed by a sentence like the following?

(S1) John has a few offenses on his record.

The existential quantifier means that John has at least one but the import of the sentence, in most conversational contexts, is that he has a number that is sufficiently high to raise certain reasonable concerns. We would still be able to get a valid argument form for the following argument:

> Anyone who has a few offenses in his record, should not be hired. John has a few offenses on his record. Therefore, John should not be hired.

Nevertheless, this seems more of a coincidence and it is, at any rate, a specific case; moreover, we have a sense that we have not rendered the deeper meaning of the sentences properly by such means. Common textbook response is to either omit such anomalies or to forswear any prospect of fixing the matter given our formal means.

Notice that we cannot do something like the following because formal languages cannot tolerate express vagueness or ambiguity; even though the meaning of the English sentence is vague (how many offenses?), the competent speaker of language can work with it. But bringing out vagueness within a formal idiom amounts to expressing ... nonsense. Hence, the following translation will not do, given that n is undefined. (The key has "Ox" to mean "x is an offense" and the ternary predicate "Hxyz" to mean "x has y on z," with John's record labeled by a constant, "r".)

$$\exists x_1 \exists x_2 \ldots \exists x_n(Ox_1 \cdot Ox_2 \cdot \ldots \cdot Ox_n \cdot Hjx_1r \cdot Hjx_2r \cdot \ldots \cdot Hjx_nr)$$

II.5 NUMERICAL PROPOSITIONS

Having the identity symbol in our artificial language allows us to symbolize numerical propositions, to which we turn now. We avail ourselves, as we have done before, of the "≠" symbol for failure of co-reference – with the same result available also by placement of the negation symbol before the subformula with the equation sign. In each case, we provide also the n-objects generalizations.

There is at least one F. $\exists xFx$

There are least two Fs. $\exists x\exists y(x \neq y \cdot Fx \cdot Fy)$

There are at least n Fs.

$\exists x_1 \exists x_2 \ldots \exists x_n(x_1 \neq x_2 \cdot x_1 \neq x_3 \cdot \ldots \cdot x_{n-1} \neq x_n \cdot Fx^1 \cdot Fx^1 \cdot \ldots \cdot Fx^n)$

There is at most one F.

$\forall x \forall y((Fx \cdot Fy) \supset x = y) \equiv \forall x \forall y(Fx \supset (Fy \supset x = y))$

There are at most two Fs.

$\forall x \forall y \forall z((Fx \cdot Fy \cdot Fz) \supset (x = y \vee x = z \vee y = z))$

There are at most n Fs.

$\forall x_1 \forall x_2 \ldots \forall x_{n+1}((Fx_1 \cdot Fx_2 \cdot \ldots \cdot Fx_{n+1}) \supset$
$(x_1 = x_2 \vee x_1 = x_3 \vee \ldots \vee x_n = x_{n+1}))$

There is exactly one F.
There is at least one F and there is at most one F.
There is a unique F.
[We may enhance our symbolic resources with a quantifier symbol for unique existence, which is defined as below: "∃!"]
$\exists!xFx \equiv \exists x(Fx \cdot \forall y(Fy \supset x = y)) \equiv \exists x\forall y(Fy \equiv x = y)$

There are exactly two Fs.
There are at least two Fs and there are at most two Fs.
$\exists x\exists y(Fx \cdot Fy \cdot \forall z(Fz \supset (z = y \vee z = x))$
There are exactly n Fs.
$\exists x_1\exists x_2...\exists x_n(Fx_1 \cdot Fx_2 \cdot ... \cdot Fx_n \cdot \forall x_{n+1}(Fx_{n+1} \supset (x_{n+1} = x_1 \vee x_{n+1} = x_2 \vee ... \vee x_{n+1} = x_n))$

All but one F is G.
$\exists!x(\sim Fx \cdot \forall y(y \neq x \supset Gy)) \equiv$
$\exists x\forall y\forall z(\sim Fx \cdot (Fy \supset y = x) \cdot (z \neq x \supset Gz))$
All but n Fs are G.
$\exists x_1...\exists x_n(Fx_1 \cdot ... \cdot Fx_n \cdot \forall y((y \neq x_1 \cdot ... \cdot y \neq x_n) \supset Gy))$

The F is the same as (or identical with) the G.
$\exists!x(Fx \cdot \exists!yGy \cdot x = y) \equiv$
$\exists x(Fx \cdot \forall y(Fy \supset y = x) \cdot \exists z(Gz \cdot \forall w(Gw \supset w = z)) \cdot x = z)$

The F is R-related to the G.
$\exists!x(Fx \cdot \exists!y(Gy \cdot Rxy)) \equiv$
$\exists x(Fx \cdot \forall y(Fy \supset y = x) \cdot \exists z(Gz \cdot \forall w(Gw \supset w = z)) \cdot Rxz)$

II.6 DEFINITE DESCRIPTIONS

The subject of definite descriptions is one of the most fascinating in the area of Logic and Philosophic Logic. Bertrand Russell's regimentation approach to the subject may be slightly more common in textbooks, while more advanced texts take pains to present additional detail. To motivate the subject we might as well make use of the celebrated topical example addressed by Russell. We have the following sentence, which, clearly, expresses a proposition insofar as it is meaningful and can be deployed correctly in declaratory or assertoric statements.

(Skf) The present King of France is bald.

Asked whether the proposition is true or false, the intuitive user of language may find herself cornered into marking it as false. There is, however, something to be said in favor of the view that this presentation of options pushes the respondent into an artificial corner. What about taking this statement to be making a proposition that fails to be true and also fails to be false given that there is no referent – no King of France exists? In the standard Propositional Logic we are examining in this text, we are compelled by an overarching Bivalence principle, encompassing other principles (presented below), which dictates that:

> *For anything to be a proposition, it has to be true or false (principle of Excluded Middle); it cannot be both true and false (principle of Non-Contradiction); it cannot take any other truth value beside one from true and false, and, hence, it cannot receive "neither true nor false" or "both true and false" as truth values.*

If the options were expanded, so that the intuitive respondent had access to the option of failure of the proposition to be either true or false, this might be the more popular answer. Even so, this answer cannot be countenanced in formal terms given our constructive means in standard Propositional Logic. What the case may be when it comes to the logic of the language – or of fragments of language – is another matter, which is fascinating but we cannot pursue it here.

Given that the proposition made by sentence Skf has to be either true or false, we wonder next as to whether "the King of France" should be considered a term – and, as such, represented by a constant symbol or a function symbol – or if this is not the case. Perhaps counterintuitively, Russell's approach, facing some opposition, refuses to take the phrase "the King of France" as a term even though this seems to be the case at first glimpse. Let us recall that we label all items in the domains, with the labels or constants having as their meanings the objects they designate. Surely, whether such an entity exists or not, the putative King of France is or would be an entity in some domain or other. Nevertheless, we cannot play the game in this way; this is a restriction incumbent on us because of the formal restrictions that stem from the way we approach the tasks of standard predicate logic. Every constant, or function, symbol must have a denotatum; otherwise, we lack meaning for our term and if we go on to deploy this symbolic resource, we basically find ourselves laying what is nonsense within our formal-linguistic framework. If this seems like a serious limitation of the formalism itself, it is to be pointed out that this anomaly can be circumvented without dismantling a formal system that we have good reasons to think successful. A stronger support, of course, is to offer that even in natural languages, like English, definite descriptions are paradigmatically not terms – even though the surface grammar may be concealing this, the logical grammar, in which we are interested, shows definite descriptions not to be terms. We will have to see how definite descriptions are to be handled.

There is a technical fix, which has not proven popular, and was conceived initially by one of the progenitors of modern logic, Gottlob Frege. We could legislate that our domain contain a non-object – which we can call the nullary object, or the unavailable object or, simply, the non-object – so that every non-referring or non-denoting descriptive term would be taken to take this non-object as its referent. In this way, we would have every term refer to something – as we must – but the way to handle terms that do not denote or refer to anything would be by making them refer to a non-object. This solution has been deemed artificial, too forced, and it is not clear that it circumvents the problem we noted above: is it meaningful to speak of non-objects?

Russell's solution is to paraphrase the sentence in the right way, so as to reveal its logical grammar which shows the definite-descriptive phrase not to be a term. This is called regimentation. It is not at all shocking – it is rather to be expected often – that the logical grammar of a sentence (or, of the proposition expressed by a sentence) is at variance with what the surface grammar shows. The issue here, however, is that the definite description is to be expressed periphrastically and not as a term; it is claimed that this is the correct to spell out the logical grammar of a definite-descriptive proposition. The item corresponding to "king of France" is to be understood as an attribute and, accordingly, we need a predicate symbol from our formal symbolic resources. If we recall that predicate symbols denote sets of objects, and given that we can have a singleton (one-member set), we can have express unique references by deploying the machinery of sets. Moreover, we can have empty sets as extensions – although, recall, we cannot have non-referring terms. Thus, a non-referring definite description will require us to take the phrase "King of France" as a logical predicate, approach it as if it had a single-member set as its extension, and avail ourselves of the semantical fiat that this extension set is empty to reach the conclusion: the proposition of Skf is to be valuated as false.

We know, from the preceding section, how to express uniqueness of possessing an attribute or unique predication. This is something we need to put to use in this case. Thus, for an evident choice of key, we have:

(Skf-Prop) $\exists!x(Kx \cdot Bx) \equiv \exists x(Kx \cdot \forall y(Ky \supset x = y) \cdot Bx)$

The regimentation can be presented analytically in the metalanguage: "There is a unique x such that x is the present King of France and x is a bald": this has the same truth conditions as, "There is an x such that x is the present King of France, and for all y, if y is the present King of France, then x and y refer to the same entity; and x is bald."

A perceptive student of logic might notice that this solution appears to incur an undesirable complication. What truth value are we to attribute to the proposition of the following sentence, and how are we to justify this valuation in relation to the valuation we presented about Skf?

(Skf_) The present King of France is not bald.

Regimentation compels the following rendering in an idiom of predicate logic like ours.

(Skf__Prop) $\exists x(Kx \cdot \forall y(Ky \supset x = y) \cdot \sim Bx)$

Since the extension of the predicate is empty, and we have conjunctions under the scope of the existential quantifier, we must also assign the truth value false to this proposition. Thus, the Skf and Skf_ are both false. They cannot be both true; they can, and indeed must, both be false. This means that they are not mutual contradictories, although one could claim, based on linguistic analysis, that they should be taken as mutual contradictories. They are contraries, but this does not alleviate the problem, if we think that the two propositions are mutual contradictories. Since we have left the concept of contradictoriness intact – we are still within the standard Propositional Logic – we seem to be confronted with an issue worth looking into.

After several years of mulling this over, Russell resolved the problem by drawing a distinction between wide scope and narrow scope of the negation symbol. Surely, we can give broader scope to the negation symbol, even making it the main connective symbol. This, of course, gives a different proposition from Skf__.

$$(\text{Skf\_\_br}) \sim \exists x(Kx \cdot \forall y(Ky \supset x = y) \cdot Bx)$$

We can contemplate formulas that are logically equivalent to Skf_br.

$$\begin{aligned}(\text{Skf\_\_br}) \sim \exists x(Kx \cdot \forall y(Ky \supset x = y) \cdot Bx) &\equiv \\ \forall x \sim (Kx \cdot \forall y(Ky \supset x = y) \cdot Bx) &\equiv \\ \forall x\, ((Kx \cdot \forall y(Ky \supset x = y)) \supset \sim Bx)&\end{aligned}$$

The broad-scope regimentation gives us a vacuously true proposition. Roughly, since there is no present King of France, the claim that "it is not the case that the present King of France is bald" is true vacuously. (Notice that the man connective symbol under the scope of the universal quantifier is now the implication symbol: hence, the vacuous true valuation for a false antecedent.) Or, it is true as the negation of the false proposition affirming that the present King of France is bald. The narrow-scope rendering reveals a false proposition: "The present King of France is not bald" is false because anything predicated of this non-existent entity fails to stick, so to speak, and the claim that this is so-and-so must be false.

We now give examples of symbolizations involving definite descriptions. A decision is needed as to whether we are to apply regimentation or treat the target phrase by means of a term symbol – specifically, an individual constants symbol. Because of this, we revisit the subject in subsequent section when we specifically address the subject of terms. (The translation examples are not necessarily true in the same model; i.e. they may not be consistent or co-satisfiable.)

Key: Cx: x is a chair.
Tx: x is a table.
Bx: x is a bed.
Yx: x is yellow.
Wxyz: x is between y and z.

(S1) There is a chair between the table and the bed.

$\exists x_1 \exists x_2(Tx_1 \cdot \forall y(Ty \supset y = x_1) \cdot Bx_2 \cdot \forall z(Bz \supset z = x_2) \cdot \exists y(Cy \cdot Wyx_1x_2))$

(S2) The only chair is to the left of the only bed.

$\exists x \exists y(Cx \cdot \forall x_1(Cx_1 \supset x_1 = x) \cdot By \cdot \forall y_1(By_1 \supset y_1 = y) \cdot Lxy)$

(S3) The only chair is yellow.

$\exists x(Cx \cdot \forall y(Cy \supset y = x) \cdot Yx)$

II.7 EXISTENTIAL PRESUPPOSITIONS

Textbooks usually mention a philosophic debate surrounding the issue of existential presuppositions which may be implied by certain, usually universally quantified, sentences of the language. For instance, we may consider the sentence shown below.

(S-universal) All American nobles are red-blooded.

Since there is and, because constitutionally enjoined, cannot be nobility in the US polity, the proposition expressed by the sentence will have to be vacuously true – as having a false antecedent.

$\forall x(Nx \supset Rx)$

This is a good opportunity to observe, as we have done in previous section, that we must create a compact predicate letter for "American noble" rather than attempt the following, wrong, symbolization.

$\forall x((Ax \cdot Nx) \supset Rx)$

One may actually have nobility status in some other country, making the antecedent true; but this is not the meaning that we are attempting to translate. If, on the other hand, we use our interpretative key to mandate that "Nx" means "x is an American noble," then the use also of the predicate Ax creates redundancy in this case.

The philosophic issue is this: if we accept the view that the competent user of language understands sentences like the one labeled S-universal above to contain existential implications – implying that there entities about which the proposition makes a generalization – then a case in which existential presuppositions fail – as in our example – should not be considered either true or false. We do not have access to the non-standard option of recognizing a third semantic value – or propositional type – which would afford us a valuation of "neither true nor false." The failure of a proposition to obtain a truth value is a disturbing notion and, in fact, it comes across as a nonsensical claim itself: by definition, proposition is meaning but we seem to be suggesting that we are dealing with nonsense; in that case, we should not accept that we are dealing with a proposition at all. Indeed, to enclose a unit of meaning within our formal analysis by means of the standard Propositional Logic, we dictate, as part of our construction instrumentalities, that any proposition is either true or false, cannot be both true and false, and cannot fail to be either true or false. We can think of this as the Bivalence Principle that characterizes standard Propositional Logic – and is extended into predicate logic. The restriction under this overriding Bivalence principle, by which a proposition must be either or false gives us the familiar principle of Excluded Middle and the restriction by which no proposition can be both true and false yields the principle of Non-Contradiction. If presupposition-impaired sentences fail to obtain a truth value, then our standard formal approach has to regard such sentences as failing to convey a propositional content. This seems to be at odds with linguistic practice, however. We could, perhaps, motivate a foray into the realm of unorthodox or non-standard (also called alternate or even "deviant") logics, but this is something that lies beyond our current pale.

II.8 TERMS AND PREDICATES: CHOICES

We could be opting for interpretative keys that assign predicate letters, instead of names, for the purpose of unique characterization of individual entities. We know how to do this based on what we have learned about definite descriptions and the regimentation or Russellian approach to the translation of such phrases. For instance, we can have the following translation, which one does not find in textbooks.

(S1) Pegasus is a flying horse.
Key: Px: x has the property of being Pegasus; x "Pegasizes."
Fx: x flies.
Hx: x is a horse.
(S1Tr1) $\exists x(Px \cdot \forall y(Py \supset y = x) \cdot Fx \cdot Hx)$

We may recall that we don't have the option of denying Pegasus' existence insofar as we cannot construct an existential predicate symbol: if we do that, affirming existence would amount to redundancy and denying existence would give us a contradiction. The sentence labeled S1 does not delve into existential or metaphysical matters, anyway. We should not think of this example as allowing us to partition non-existent things from things that exist: so, we are not dealing with a model that, including Pegasus, represents a possible approach to metaphysical issues about what exists and what does not. This is because any model commits us wholesale and automatically to the existence of its objects. This is to be understood on the grounds that metaphysical issues – deeper reflections on existence and on the kinds of things that can exist – are locked out, and are not allowed to surface, within our formalism. Our present inquiry in this section is a narrower one. We showed above how the translation can be carried by means of a dedicated predicate letter in the signature of our model. Of course, we can also try a translation in which the object Pegasus is labeled by means of a constant, "p".

(S1Tr2) $Fp \cdot Hp$

Although it is not the practice to translate as in S1Tr1, we might wonder if it really makes any difference. The translation in S1Tr1 would come handy in solving certain problems – which do not concern us here – if we worked with the formal system of Aristotelian or Categorical Logic, which we do not study in the present text. The categorical formula types of the system include the form "All As are Bs" but do not provide for naming entities. (Naming is a deeply fascinating subject on its own right and it pops up often in the philosophy of logic. It is not surprising if Aristotle had some notions about names, which prevented him from treating names the way we have done in setting up predicate logic – taking the meanings of names to be merely their referent objects. A view that may be intuitively more appealing is that names are not rendered meaningful simply the way labels or tags are meaningful; notice, however, that our predicate formalism actually takes this approach and capitalizes on it to good effect.) By means of the translation we saw above, we can fit a sentence like "Socrates is wise" by rendering it as "All Socratizing things are wise things," – with the uniqueness caveat remaining unexpressed.

Can we validly derive from the formula in S1Tr1 that Pegasus is a horse? How would we symbolize, with the P-predicate, the proposition we would want to be able to derive validly as a conclusion?

(S1Concl) $\exists x(Px \cdot \forall y(Py \supset y = x) \cdot Hx)$

We can derive this formula validly from S1Tr1; but we can adduce against this option that it does not seem to capture the deeper structure of the proposition – with names being terms and the logical predicates attaching to them. Added unnecessary complexity of the formulation also counts against this option.

(S2) A witch stole the ring.

Here we face a challenge as to whether we should use a predicate interpreted as "ring___" or if we should rather label or name the specific ring. Here are the two translations – for an evident choice of interpretative key.

(S2Tr1) $\exists x(Wx \cdot \exists y(Ry \cdot \forall z(Rz \supset z = y) \cdot Sxy))$
(S2Tr2) $\exists x(Wx \cdot Sxr)$

It is difficult to find principled or deeply based objections to either translation but there may be broader objectives that can guide our choice. For instance, we might be interested in deriving the proposition of the sentence:

(S2Concl) A witch stole exactly one ring.

We show the translation of the conclusion, which is the same regardless of whether we take our interpretative bearings from S2Tr1 or S2Tr2.

(S2ConclTr) $\exists x(Rx \cdot \forall y(Ry \supset y = x) \cdot \exists z(Wz \cdot Szx))$

Since this is a valid derivation from S2Tr1 but not from S2Tr2, we may then reasonably opt for the former translation.

II.9 DISAMBIGUATIONS

Natural languages present ambiguity – as well as vagueness, with which we do not deal in the present text as it would require forays into non-standard logics. Ambiguity may be desirable, as in creative writing. We have ambiguity when the competent user of language may reasonably understand one or the other of at least two different meanings of the sentence (which may be because of more than one possible meanings of words or phrases or even because of grammatical construction that can be rearranged in different ways.) On the other hand, formal languages do not permit ambiguity – with ambiguous expressions ruled out by the grammatical arrangements and, in the semantics, by the conditions on models. Thus, it is not logically possible to translate an ambiguous sentence by an ambiguous formula; this is undesirable, anyway, given that ambiguity that is genuine or irremovable (and does not serve an extra-logical linguistic purpose) is a pathology.

We implement disambiguation when we provide all the translations for a given sentence of the natural language, with each translation corresponding to each plausible context of interpretations. Given the ambiguity, the sentence could be expressing one of a number of distinct propositions, with each such proposition generating a specific context. Each such proposition may be called a disambiguation or disambiguating proposition of the given sentence. (Notice that it is the sentence that needs to be disambiguated while a proposition is always free of ambiguity. The ambiguous sentence may be taken competently to be expressing any one of various different propositions. Those propositions are distinct in the sense that they have different truth conditions – they are not logically equivalent with each other.) Obviously, disambiguating propositions may – although they do not have to – be mutually inconsistent.

Disambiguating involves us more often than not in discerning alternative possibilities regarding interplays between the scopes of quantifiers and other symbols. In other words, scope ambiguity – a bane of natural language – is paramount in examples of this sort.

As examples, we may consider the following.

> (G) All that glitters is not gold.
> Key: Gx: x glitters; Lx: x is gold.
> (DisambG1) $\sim \forall x(Gx \supset Lx) \equiv \exists x(Gx \cdot \sim Lx)$
> (DisambG2) $\forall x \sim (Gx \supset Lx) \equiv \forall x(Gx \cdot \sim Lx)$

The natural language expresses the proposition in such a grammatical formation that the ambiguity arises ineluctably and the ministrations of logical analysis are needed urgently. The first disambiguation seems to be the proper one in the sense that, most commonly, in linguistic contexts, this is the intended interpretation. The second disambiguation comes across as unnatural for formalist reasons – as the universally quantified formula has a conjunction rather than implication within its scope. This, however, cannot be used as the reason for rejecting it when we discuss translations from language. An interesting problem here is that the user of language would be intending the second disambiguation under conditions in which the meaning is intensional – something along the lines of "for all things, it is not necessarily the case that if they glitter they are gold." This rendering would compel us to take the negation as having a different meaning from the familiar extensional meaning we find in Propositional Logic. This cannot be ruled out even though it is unorthodox from the standpoint of the standard formalism. As it is, the first disambiguation seems to be the most common one but, in natural language discourse, context is the ultimate relevant element in determining meaning.

(L) Mark likes himself and so does Jill.
Key: m: Mark; Lxy: x likes y; j = Jill.
(DisambL1) Lmm · Ljj
(DisambL2) Lmm · Ljm

In this case, we face a case of ambiguity that is not a matter of scope; so, this is certainly possible.

(R) Everyone likes someone.
Key: Lxy: x likes y.
(DisambR1) ∀x∃yLxy
(DisambR2) ∃x∀yLyx

This staple example of textbooks is ambiguous as conveyed by a sentence of the natural language. Of course, except in rare contexts, it is most likely the first interpretation or disambiguation that is intended by the user of language. Nevertheless, the prospect of ambiguity, depending on linguistic context, is also present but, of course, this never arises within a formal language.

Sometimes disambiguation is not possible, which impugns our formal system's aptitude for capturing all the nuances of language but not the mechanics of the formal system itself which – like, for instance, the Euclidean geometry – is beyond reproach when taken simply as a formal system. A significant example of a disambiguation we cannot perform based on the resources of the predicate logic, is between "any" and "all" which, in certain contexts, can only be translated by the same formula even though they have different truth conditions.

For example, we do not run into ambiguity in the following example because, in this case, "any" and "all" have the same logical meaning – i.e. truth conditions.

(S1) Any soldier is brave.
(S2) All soldiers are brave.
$\forall x(Sx \supset Bx)$

In the following example, however, the logical meanings of the sentences are not identical; this owes to the participation of "any" and "all" in the compositions of the sentences (and "any" and "all" are certainly logical words) and, yet, this difference cannot be captured by means of appropriately different symbolizations.

(S3) Any soldier might fail the test.
(S4) All soldiers might fail the test.

It might appear at first that the culprit is the non-truth-functional particle "might" but we have a way around this particular problem by means of constructing a predicate letter that is interpreted as follows:

Fx: x is liable to fail the test.

Yet, the translations of S3 and S4 are identical:

$\forall x(Sx \supset Fx)$

Another, somewhat more exotic, example shows that we cannot use predicate symbolic resources to disambiguate between two different notions which medieval logicians called "de dicto and "de re." This may be considered inevitable since the two notions are, by their definitions, indebted to specific interplays between modal contexts and what objects lie within their scopes. We give an example:

(S5) Someone will travel to Alpha Centauri sometime in the future.

The ambiguity incurred by the sentence of the natural language is between two different possible notions:

> De Re: There is someone at the present (which is determined indexically) who will travel to Alpha Centauri in the future
>
> De Dicto: There is someone who is in the future (indexically determined, again, from a present standpoint) who will then travel to Alpha Centauri.

We can construct a predicate letter with the appropriate interpretation, and we can also use predicate symbols to capture references to future times, but we cannot disambiguate between the two notions.

> Key: Tx: x is a time; Fx: x is future; Px: x is a person;
> Rxab: x travels to Alpha Centauri at time b.
> $\exists x(Px \cdot \exists y(Ty \cdot Fy \cdot Rxay))$

We might try to construct a predicate letter Lxy interpreted as "x is alive at y" but the resultant symbolization will not permit us to make certain important inferences in advanced logic regarding the relationship between de dicto and de re notions.

II.10 EXAMPLES

- Recapitulation with Examples

I. One-place predicates (with identity).
No one is an F/ no one Fs: $\sim \exists xFx$
Not everyone is an F: $\sim \forall xFx$
Someone is an F: $\exists xFx$
Everyone is an F: $\forall xFx$

All F are G: $\forall x(Fx \supset Gx)$ ------- [all/everyone goes together with <u>if-then</u>]
Some F are G: $\exists x(Fx \cdot Gx)$ -------- [something/someone goes together with <u>and</u>]
No F are G: $\sim \exists x(Fx \cdot Gx)$
Some F are not G: $\exists x(Fx \cdot \sim Gx)$
Not all F are G: $\sim \forall x(Fx \supset Gx)$
Remember also the equivalences:
$\forall xFx \equiv \sim \exists x \sim Fx$ [everyone Fs = no one does not F]
$\sim \forall xFx \equiv \exists x \sim Fx$ [not everyone Fs = someone does not F]
$\sim \exists xFx \equiv \forall x \sim Fx$ [not anyone (no one) Fs = everyone does not F]
$\exists xFx \equiv \sim \forall x \sim Fx$ [someone Fs = not the case that everyone does not F]

Key: Px: x is a politician./Cx: x is corrupt./Jx: x is going to jail./j: John./m: Mary.
---- Our domain has only persons.
If John is a politician, then he is corrupt: $Pj \supset Cj$
Mary is not a politician but she is corrupt: $\sim Pm \cdot Cm$
Either Mary is a politician or there is no politician (no one is a politician): $Pm \vee \sim \exists xPx$
Given that all politicians are corrupt, then if John is a politician he is corrupt:
$\forall x(Px \supset Cx) \supset (Pj \supset Cj)$
Every politician is either corrupt or is not John: $\forall x(Px \supset (Cx \vee x \neq j))$
Not every politician is corrupt: $\sim \forall x(Px \supset Cx)$
Some corrupt politician is going to jail: $\exists x(Px \cdot Cx \cdot Jx)$
If no corrupt politician goes to jail, then every politician who does not go to jail is corrupt: $\sim \exists x(Px \cdot Cx \cdot Jx) \supset \forall x((Px \cdot \sim Jx) \supset Cx)$
John and Mary are politicians if everyone who is corrupt is a politician: [notice that what follows "if" goes before the horseshoe.]
$\forall x(Cx \supset Px) \supset (Pj \cdot Pm)$
There are exactly two corrupt politicians and they are John and Mary:
$\exists x\exists y(Px \cdot Py \cdot Cx \cdot Cy \cdot x \neq y \cdot x = j \cdot y = m)$

II. Two-place predicates (with identity).
Everyone is R-related to everyone: $\forall x \forall y Rxy$
Everyone is R-related to himself/herself: $\forall x Rxx$
Someone is R-related to someone: $\exists x \exists y Rxy$
Someone is R-related to someone other than himself/herself: $\exists x \exists y(Rxy \cdot x \neq y)$
Everyone is R-related to someone other than himself/herself: $\forall x \exists y(Rxy \supset x \neq y)$
No one is R-related to anyone: $\sim \exists x \sim \exists y Rxy$
Someone is R-related to no one: $\exists x \sim \forall y Rxy$
Everyone is related to no one: $\forall x \sim \exists y Rxy$
Not everyone is R-related to someone: $\sim \forall x \exists y Rxy$
==============================
Key: Px: x is a politician./Cx: x is corrupt./Jx: x is going to jail./ Bxy: x bribes y/ j: John./m: Mary.
-------------Our domain has only persons.
If someone gives bribes to John, it follows that John is corrupt but he is not going to jail:
This one is ambiguous: there are two possibilities:
$\exists x Bxj \supset (Cj \cdot \sim Jj)$ and
$\exists x(Bxj \supset Cj) \cdot \sim Jj)$
Anyone who gives bribes to Mary is corrupt: $\forall x(Bxm \supset Cx)$
Either John bribes Mary or no one does: $Bjm \vee \sim \exists x Bxm$
If someone bribes everyone, then no one goes to jail: $\exists x \forall y Bxy \supset \sim (\exists x)Jx$
If Mary bribes John and John is not corrupt, then John goes to jail: $(Bmj \cdot \sim Cj) \supset Jj$
Everyone who bribes some politician is corrupt: $\forall x)(\exists y(Py \cdot Bxy) \supset Cx)$
[Notice what happened here: the existential quantifier phrase has to go inside the parenthesis. We are saying: for all x – if there is some y, so that ---- - then x is corrupt.]
Anyone other than John bribes some corrupt politician: $\forall x(x \neq j \supset \exists y(Py \cdot Cy \cdot Bxy))$

No corrupt politician, who is not Mary, goes to jail: $\sim \exists x(Px \cdot Cx \cdot x \neq m \cdot Jx)$
If Mary bribes John, then either John is a polician or Mary is going to jail:
$Bmj \supset (Pj \vee Jm)$
There are at least two people who bribe some corrupt politician:
$\exists x \exists y(x \neq y \cdot \exists z(Pz \cdot Cz \cdot Bxz \cdot Byz))$
==

III. Expressing uniqueness --- <u>*the*</u> F.
The King of England is bald.
Key: Kxy: x is a King of y./ Bx: x is bald./e: England./
We translate: there is at least one x that is King of England, and every y - if y is a King of England, then y is identical with x - and x is bald. {This is how we capture uniqueness.}

$\exists x(Kxe \cdot \forall y(Kye \supset x = y) \cdot Bx)$

Equivalently:

$\exists x \forall y((Kye \equiv y = x) \cdot Bx)$

- Jack and Jill are not husband and wife.

  Key: a: Jack; b: Jill.

  Hxy: x is the husband of y.

  Wxy: x is the wife of y.

  $\sim Hab \cdot \sim Wba$

  This is the most likely translation across possible linguistic contexts. It is possible, however, that in some context or other, something else is meant. Thus, we are facing a disambiguation challenge. A disambiguating option, given below, would have the sentence stating that Jack and Jill are not husband and wife, respectively, although not in the sense that they are not each other's husband or wife.

  $\sim \exists xHax \cdot \sim \exists yWby$

Of course, given the legal permissions in a society like ours, the following is analytically true – but not because of its logical form. We speak elsewhere of the possibility of adding meaning postulates which elevate, for a given model, contingent truths like this one to the status of honorary truths in the model – so to speak. Still, a model can be constructed in which the formula is false; this shows that this is not a logical truth – it is not a logically necessary truth by virtue of the formula's logical form.

$$Hab \equiv Wba$$

- Anyone who admires himself or herself admires at least one person.

$$\forall x(Axx \supset \exists yAxy)$$

It is not stated that the admired person in the consequent is someone other than the one who admires. In some contexts, the point might be that one unique person is being admired – who will have to be the one who also admires himself or herself. This is unlikely.

- The Statue of Liberty is in New Jersey.

Key: s: the Statue of Liberty; n: New Jersey.

Lxy: x is located at y.

Lsn

Another possible translation is by use of the machinery for definite descriptions. Taking "being the Statue of Liberty" to be a logical predicate (Sx), we need to express uniqueness (which is ensured automatically when a term, an individual constant or a function, is used in the translation since terms cannot refer to more than one object – although they can co-refer.)

$$\exists x(Sx \cdot \forall y(Sy \supset y = x) \cdot Lxn)$$

- The alleged serial killer is on trial.

Key: Kx: x is an alleged serial killer.

Tx: x is on trial.

$\exists x(Kx \cdot \forall y(Ky \supset y = x) \cdot Tx)$

We cannot decompose into the grammatical adjectival qualifiers – alleged, serial and killer – for reasons we explained in this section.

Compare:

An alleged serial killer is on trial.

$\exists x(Kx \cdot Tx)$

At least one person is on trial in the trial of an alleged serial killer. –Notice that, now, we need to shift to an unrestricted domain (whereas we had persons-domain before) and we have, accordingly, to use symbols for being a trial and for being a person.

Key: Tx: x is a trial.

Px: x is a person.

Oxy: x is on (trial) y.

$\exists x \exists y(Ty \cdot \forall z(Tz \supset z = y) \cdot Px \cdot Oxy)$

- We provide here multiple examples of translations into the formal language PRED. We translate not exactly from idiomatic English as a natural language but from formulaic sentences that express – rather artificially – archetypal propositional forms. This list is not and cannot possibly be and is not exhaustive with respect to interesting permutations of first-order logical phrases. As it is, it can serve as a reference guide for translations. In some instances, paradigmatically, we show steps in a translation process. It is rare that we need to specify a key, as you will ascertain. We often show logically equivalent formulas to stimulate reflection and advance facility with the notions of predicate logic.

a. At least one F is not R-related to all Gs.

$\exists x(Fx \cdot \sim \forall y(Gy \cdot Rxy)) \equiv \exists x(Fx \cdot \exists y \sim (Gy \cdot Rxy)) \equiv \exists x(Fx \cdot \exists y\, (\sim Gy \vee \sim Rxy)) \equiv \exists x(Fx \cdot \exists y\, (Gy \supset \sim Rxy)) \equiv \exists x(Fx \cdot \exists y\, (Rxy \supset \sim Gy))$

b. An F is R-related to a G.

Disambiguations:

$\exists x(Fx \cdot \exists y(Gy \cdot Rxy))$

$\exists !x(Fx \cdot \exists y(Gy \cdot Rxy))$

$\exists x(Fx \cdot \exists !y(Gy \cdot Rxy))$

$\exists !x(Fx \cdot \exists !y(Gy \cdot Rxy))$

Recall that:

$\exists !xFx \equiv \exists x(Fx \cdot \forall y(Fy \supset x = y)) \equiv \exists x \forall y(Fy \equiv x = y)$

c. Whenever it Fs, it also Gs.

"it" must be labled by a constant: a

Fxy: x Fs at time-period y.

Tx: x is a time-period.

$\forall x((Tx \cdot Fax) \supset Gax) \equiv \forall x(Tx \supset (Fax \supset Gax))$

d. It G-ed in the past but it will H in the future.

Key as in c; added key: Px: x is a past time-period; Ux: x is a future time-period.

$\exists x(Tx \cdot Px \cdot Gax) \cdot \exists y(Ty \cdot Uy \cdot Hay)$

e. The Fs will, one day, be R-related to the Gs.

Ox: x is an object.

Dx: x is a day.

Ux: x is future.

$\forall x(Fx \supset \exists y(Dy \cdot Uy \cdot Rxy))$

i. Only Fs are Gs.

For all x, x Gs only if x Fs.

We recall from Propositional Logic translations that the necessary condition – introduced by "only if" – is in the position of the consequent of a conditional or implicative formula.

$\forall x(Gx \supset Fx)$

j. Every F's G is also an H's G.

Bxy: y belongs to x.

[note: we could set as key: Bxy: x belongs to y. Do we have any incentive to specify the key the way we have done?]

Disambiguations:

$\forall x \forall y((Fx \cdot Gy \cdot Bxy) \supset \exists z(Hz \cdot Bzy))$

--all Gs that are Fs belong to some H.

$\forall x \exists! y((Fx \cdot Gy \cdot Bxy) \supset \exists z(Hz \cdot Bzy))$

--a unique G is F's.

$\forall x \forall y((Fx \cdot Gy \cdot Bxy) \supset \exists! z(Hz \cdot Bzy))$

--any F's G belongs to a unique H.

$\forall x \exists! y((Fx \cdot Gy \cdot Bxy) \supset \exists! z(Hz \cdot Bzy))$

--both the F's G and the H to which it belongs are unique.

$\forall x(\exists y(Fx \cdot Gy \cdot Bxy) \supset \exists z(Hz \cdot Bzy))$

--if there is some F's G....

$\forall x(\exists! y(Fx \cdot Gy \cdot Bxy) \supset \exists z(Hz \cdot Bzy))$

$\forall x(\exists y(Fx \cdot Gy \cdot Bxy) \supset \exists! z(Hz \cdot Bzy))$

$\forall x(\exists! y(Fx \cdot Gy \cdot Bxy) \supset \exists! z(Hz \cdot Bzy))$

To develop the uniqueness clause, as we do with regimented definite descriptions, we show the second disambiguation as example:

$\forall x \exists! y((Fx \cdot Gy \cdot Bxy) \supset \exists z(Hz \cdot Bzy)) \equiv$
$\forall x \exists y((Fx \cdot Gy \cdot \forall w(Gw \supset y = w)) \supset \exists z(Hz \cdot Bzy))$

-- Do the expansions for the rests as an exercise.

k. Every F that is R-related to a G is also S-related to an H.

$\forall x(Fx \supset (\exists y((Gy \cdot Rxy) \supset \exists z(Hz \cdot Sxz))))$

Spell out disambiguations as an exercise. The indefinite article "a" may or may not be referring to a unique referent.

l. All those that are decidedly F are G if they are not R-related to some H or other.

"decidedly" cannot be rendered by truth-functional means.

$$\forall x(Fx \supset (\sim \exists y(Hy \cdot Rxy) \supset Gx)) \equiv$$
$$\forall x((Fx \cdot \sim \exists y(Hy \cdot Rxy)) \supset Gx) \equiv$$
$$\forall x((Fx \cdot \forall y(\sim Hy \vee \sim Rxy)) \supset Gx)$$

m. Fs that are R-related to each other, are related to some G.

$$\forall x \forall y((x \neq y \cdot Rxy) \supset \exists z(Gz \cdot Rxz \cdot Ryz))$$

Is it usually implied that the G is other than the x and y? Could that be implied in some context?

We could produce the following disambiguating option:

$$\forall x \forall y((x \neq y \cdot Rxy) \supset$$
$$\exists z(z \neq x \cdot z \neq y \cdot Gz \cdot Rxz \cdot Ryz))$$

n. Every F that is R-related to something else is not R-related to itself.

$$\forall x((Fx \cdot \exists y(y \neq x \cdot Rxy) \supset \sim Rxx)$$

Could it be implied, in some context, that the "other" thing is also a non-F?

$$\forall x((Fx \cdot \exists y(y \neq x \cdot \sim Fy \cdot Rxy) \supset \sim Rxx)$$

o. None of the Fs are R-related to anything but a G.

$$\sim \exists x(Fx \cdot \exists y(Rxy \cdot \sim Gy))$$

p. The F is one of two G.

$$\exists! x(Fx \cdot \exists y \exists z(Gy \cdot Gz \cdot y \neq z \cdot x = y \vee x = z))$$

q. The Gs that are R-related to the Fs are R-related to each other.

$\forall x \forall y((Gx \cdot Gy \cdot \forall z(Fz \supset (Rxy \cdot Ryz)))$
$\supset (Rxy \cdot Ryx))$

r. Sometimes a Fs and sometimes it Gs.

$\exists x(Tx \cdot Fax) \cdot \exists y(Ty \cdot Gay)$
$\exists x((Tx \cdot Fax) \cdot \exists y(Ty \cdot y \neq x \cdot Gay))$

s. If a exists, then it is R-related to a b that is a G.

$\exists x(x = a \supset \exists y(y = b \cdot Gy \cdot Rxy))$
$\exists x(x = a \supset \forall y((y = b \cdot Gy) \supset Rxy))$
$\exists x(x = a \supset \exists !y(y = b \cdot Gy \cdot Rxy))$

t. If someone Fs, then someone is R-related to him/her.
Caution is needed. We should not translate as follows:
(!) $\exists xFx \supset \exists yRyx$ y is free in the consequent/ hence, no proposition is expressed
Nor should we translate as:
(!) $\exists x(Fx \supset \exists yRyx)$
This means: "there is someone, such that someone is R-related to him/her."
The translation should be:
$\forall x(Fx \supset \exists yRyx)$
The point is that anyone who Fs is R-related to someone.
In contrast, suppose that we have:
If someone who Fs exists, then someone is R-related to him/her.
Analytically: for anyone, such that he/she Fs and exists, then someone is R-related to him/her.
We express existence of an object labeled by "a" as follows:
$\exists x(x = a)$

Returning to our example – and also bearing in mind that we are not told that the person R-related to the initial person must be a different person – we continue:

$$\forall x(\exists y(x = y \cdot Fx) \supset \exists zRzx)$$

The equivalent prenex form – and equivalent formulas – are shown below:

$$\forall x \forall y \exists z((x = y \cdot Fx) \supset Rzx) \equiv$$
$$\forall x \forall y \exists z(x = y \supset (Fx \supset Rzx))$$

Disambiguating for the option that has a different person related to the person who Fs, we have:

$$\forall x(\exists y(x = y \cdot Fx) \supset \exists z(z \neq x \cdot Rzx))$$

u. John's bark is worse than his bite.

We need the symbolic resources of second-order logical formalism because the relational (binary) predicate "--- is worse than ___" in this sentence takes as inputs (also called arguments) first-order predicates. We also need quantifier symbols ranging over first-order predicates. But we will need to make do, as much as we can, with first-order predicate logic resources. Here is a suggestion. We may wonder about what is lost: this should mean that some argument that are intuitively apprehended as valid by the competent user of language may be first-order invalid when symbolized along the lines indicated here. Such an argument should, presumably, be second-order valid when translated by using the resources of second-order logic. We run into such phenomena with respect to argument forms that are propositionally invalid but first-order valid; this happens also when we ascend to extending propositional or predicate logic to modal logics.

KEY: Bxy: x bites at y.

Rxy: x barks at y.
Tx: x is a time period.
Wxyzw: what x does at y is worse than what z does at w.
$\forall x \forall y((Tx \cdot Ty \cdot Rjx \cdot Bjy) \supset Wjxjy)$

- Now we show translations into PRED – an idiom of predicate logic – of propositions often expressed by metalinguistic sentences in advanced texts dealing with Modal Logic. More broadly, the propositions define properties of properties, but do not require resources of higher-order logic for their expression in this context. In this way, we define such properties as reflexivity, symmetry, and so on.

R is Reflexive: $\forall x Rxx$
R is Irreflexive: $\forall x \sim Rxx$
R is Non-Reflexive: $\exists x \sim Rxx$
R is Symmetric: $\forall x \forall y(Rxy \supset Ryx)$
R is Asymmetric: $\sim \forall x \forall y(Rxy \supset Ryx) \equiv \exists x \exists y(Rxy \cdot) \sim Ryx)$
R is Non-Symmetric: $\exists x \exists y(Rxy \cdot \sim Ryx)$
R is Anti-Symmetric: $\forall x \forall y((Rxy \cdot Ryx) \supset x = y)$
R is Transitive: $\forall x \forall y \forall z((Rxy \cdot Ryz) \supset Rxz)$
R is Intransitive: $\forall x \forall y \forall z((Rxy \cdot Ryz) \supset \sim Rxz)$
R is Non-Transitive: $\sim \forall x \forall y \forall z((Rxy \cdot Ryz) \supset Rxz) \equiv \exists x \exists y \exists z(Rxy \cdot Ryz \cdot \sim Rxz)$
R is Serial: $\forall x \exists y Rxy$
R is Euclidean: $\forall x \forall y \exists z((Rxy \cdot Rxz) \supset (Ryz \vee Rzy))$
R is Dense: $\forall x \forall y \exists z((Rxy \cdot Ryz) \supset (Rxz \cdot Rzy))$
R is Convergent: $\forall x \forall y \forall z \exists w((Rxy \cdot Rxz) \supset (Ryw \cdot Rzw))$
R is Connected: $\forall x \forall y(Rxy \vee Ryx)$
R is One-Step Dead-End: $\forall x \forall y(Rxy \supset \forall z(Ryz \supset z = y))$

- Natural languages teem with idiomatic expressions that can be understood by the competent user of the language. A term that has been coined to capture those aspects of the use of language that fall outside the purview of formal logic is "pragmatics." There is disagreement, however, on how much of what is presumed to belong to pragmatics, in this sense, may be brought under some appropriately extended and resourceful formal system. Without entering into such advanced thickets, we will show here examples of how certain idiomatic expressions may be negotiated when it comes to translating into some predicate idiom like ours. We will also take the unusual, but available, liberty of adding to our formal resources. Strictly speaking, we are extending PRED by adding those symbolic resources for which we also need to specify grammar and semantic or truth conditions. We add two connective symbols. One of them presents certain theoretical problems, which do not concern us here; we may think of it as the "absurdity symbol" and it is treated as a logical constant or zero-place function. This is a mathematical artifice, rather elegant too, by which a connective is considered to be a zero-place or zero-input connective when it defined as a fixed truth value – in this case the truth value false. We could not write out a truth table definition for this connective. As you can see from its definition – which we carry out by means of an equivalential formula – we don't have to have this symbol since it is inter-substitutable (without altering the truth value of the compound) with the formula by which we define it below. The other connective symbol we add is binary and it has the same truth table as negated material equivalence. This symbol is available for symbolization of exclusive disjunction because it turns out – for the standard Propositional Logic – that negated equivalence (or disequivalence, as it is also called) has the same logical meaning (truth conditions, truth table) with the connective we would construct for excusive disjunction. (For exclusive disjunction, we require, by definition, that exactly one – definitely one and only one – of the two input values is true. The available combinations, then, are those that have true for the first and false for the second input or false for the second and true for the first: but this is the truth table for negated equivalence.)

$\bot \equiv (p \cdot \sim p)$
$(p \not\equiv q) \equiv \sim (p \equiv q)$

Translations:

If John is ready for the exam, then I am a monkey's uncle.

The intent is to state that a logical absurdity is implied by taking the problematic proposition as antecedent. Indeed, if you do the truth table computations, you can see that this implicational formula is logically equivalent with the negated antecedent.

$Rj \supset \bot$

Either John or Mary is president of the Speluncean Club.

In most standard contexts, the meaning of the connective is exclusive disjunction; it has to be exactly one president unless, more rarely, the office of the presidency can be shared (collective presidency is permitted.)

$Pj \not\equiv Pm$

We may wonder how we can show more context. Here is suggestion.

Key: Cx: x is a club; s: the Speluncean Club; Px: x is a person; Rx: x is President.

$\exists!x(Cx \cdot x = s \cdot \exists!y(Py \cdot Ry \cdot y = j \not\equiv y = m))$

We carry out transformations to logically equivalent expressions. It is a good idea to concentrate on studying this subject and acquiring some facility.

$\exists!x(Cx \cdot x = s \cdot \exists!y(Py \cdot Ry \cdot (y = j \not\equiv y = m))) \equiv$
$\exists x\forall y((Cy \equiv y = x) \cdot x = s \cdot \exists z\forall w((Pw \cdot Rw) \equiv w = z) \cdot (z = j \not\equiv z = m)))$

Other options are available for the translation. We could use function symbols, for instance. We could also capture "the Speluncean Club" by means of a constant but it is arguable that this leaves out some relevant detail about the attribute of being-a-Club. Here are some other options for the translation. You can figure out the key that is used in each case.

$p(s) = j \not\equiv p(s) = m$
$\exists!x(Sx \cdot (p(x) = j \not\equiv p(x) = m)) \equiv$
$\exists x\forall y((Sy \equiv y = x) \cdot (p(x) = j \not\equiv p(x) = m))$

"Sometimes a cigar is just a cigar."

This sentence is attributed to Sigmund Freud. There is an idiomatic twist in this sentence, in a broad sense. There is a tendency to translate such sentences (compare "what will be will be") as expressing trivial or tautological propositions of the logical form of self-implication. Nevertheless, it does not appear to be the speaker's intention to state something that is trivial or logically self-evident. Occasions may arise in which a speaker utters a trivial statement – perhaps to mock the lack of intelligence regarding what is logically self-evident on the part of her interlocutors. Freud's point is that slips of the tongue, and other so-called parapraxes, do not occur always or inevitably. The challenge for us is to capture sufficient structure in our translation and this, at a minimum, presupposes that we steer clear of misrepresenting the logical form of the stated proposition. Here is a suggested translation.

KEY: Tx: x is a time period.
C_1xy: x is stated to be a cigar at time y.
C_2xy: x is meant to be a cigar, as talked about, at time y.

$\exists x(Tx \cdot \forall y(C_1yx \supset C_2yx)) \equiv \sim \forall x(Tx \supset \forall y(C_1yx \cdot \sim C_1yx))$

II.11 EXERCISES

(1) Translate the propositions expressed by the given meaningful English sentences by using the given key – lexicon.

j: John/m: Mary/Lxy: x likes y/Gxyz: x gives y to z/Px: x is a person/Ty: y is a thing

a. John likes Mary but he doesn't give her anything. [--there is no thing that he gives her.]
b. Mary likes herself but she doesn't give herself anything.
c. John does not like himself and anything he gives he gives to Mary.
d. Either Mary gives something to John or John does not like himself.
e. Mary likes John if John doesn't like himself.
f. If John likes Mary if she gives him something, then John doesn't like himself.
g. Only Mary likes John without giving him anything. [For all x, if x likes John and there is no y that is a thing and x gives to y, then x is identical with Mary.]
h. Mary likes John if John likes everyone.
i. Mary likes John if John likes everyone except himself.
j. John likes Mary if and only if Mary gives him something.
k. Only if Mary does not like John does John give her something. [Recall: only if φ, ψ: φ is the consequent!]
l. Unless John and Mary do not like each other, they don't give anything to each other. [--- there is nothing that John gives to Mary and there is nothing that Mary gives to John.] [How is "unless" to be treated? Options include negated conjunction and negated equivalence, which is the same as exclusive disjunction.]
m. Anyone who likes Mary without liking John gives nothing to John and gives something to Mary. [For all x, if x is a person --]
n. If John gives anything to Mary, then Mary does not like John.

o. The only person who likes John is Mary. [for all x, if x likes John, then x is identical with Mary.]
p. The only person who likes John is John.
q. Everyone who likes John also likes Mary.
r. No one who gives anything to John likes John.
s. Not everyone gives something both to John and to Mary. [Ambiguity: the same thing or different things are given to John and Mary? – we must render separate translations to disambiguate.]

(2) Make you own key and translate the propositions expressed by the following English sentences into our formal language PRED. Disambiguate, as needed, by presenting all the available disambiguating options.

a. Either you win or you lose. [Consider that this is a generalization.]
b. If you don't win, then someone else does.
c. No one dislikes everyone.
d. The Biology teacher is liked by everyone.
e. The Philosophy teacher is liked less than the Biology teacher.
f. John ate all the cakes.
g. John ate some cakes. [What problem are we facing here?]
h. If students are happy, so is John. [Do we need disambiguation?]
i. Jack and Jill are unhappy when they go to parties. [Is the point that they have to be both present?]
j. Jill is a bright businesswoman.
k. John is the alleged embezzler. [Problem?]
l. Anyone who admires himself admires at least one person.
m. Someone gave something to someone else without being given anything in return. [Unrestricted domain is needed.]
n. Only if everyone is happy is everyone safe.
o. Unless everyone is safe, no one is happy. [Recall issues surrounding the translation of "unless" from propositional translations.]
p. The suspected serial killer is on the loose.
q. If a suspected serial killer is on the loose, then no one is safe.

(3) Translate from PRED into English, striving for a more naturally idiomatic translation in the target language - which is English. You should not have ambiguity challenges in the material you are translating. [Why?] Can you always make sure that you avoid ambiguity in the English translation (while you are also observing the standard of choosing an idiomatically natural translation into English?)

KEY: Rxy: x respects y.
r(x): the elected representative of x's district.

a. $\exists x \exists y Rxy$
b. $\exists x(\exists y Rxy \vee \forall z Rxz)$
c. $\forall y \exists z(Ryz \supset Rzy)$
d. $\exists y(\exists x Rxy \supset \sim \forall z Rzy)$
e. $\forall x Rxx \supset \exists y \exists z \sim Ryz$
f. $\forall x \forall y((x \neq y \cdot Rxy) \supset Ryx)$
g. $\exists x(\exists y(Rxy \cdot x \neq y) \equiv \forall z Rzy)$
h. $\exists x \exists y \forall z((Rxy \cdot Ryx \cdot x \neq y) \supset (Rxz \vee Ryz))$
i. $\sim \forall x(Rxx \supset \exists y(y \neq x \vee Ryx))$
j. $\forall x \forall y(x = y \equiv (Rxy \equiv Rxx))$
k. $\exists x(Rxr(x) \supset r(x) = x)$
l. $\forall x \exists ! y(r(x) = y)$
m. $\exists x \sim Rxr(x) \supset \forall x \forall y(y = r(x) \supset \sim Rxy)$

(4) Consider the following meaningful sentences of English. In all of them the verb "to be" appears in some grammatical permutation. As a matter of logical grammar, however, there are distinct meanings of the verb "to be": <u>copula</u> connecting logical predicate to the subject of which the property is predicated; <u>existential affirmation</u>; <u>identity</u> (co-reference.) Identify which logical meaning the verb has in the proposition expressed by each one of the sentences. The lumping together of all these disambiguated logical meanings of this verb has led to disastrous results in the history of philosophy - discovering problems and paradoxes because of being misled by the surface grammar and ignoring the multiplicity of meanings when the logical

grammar is examined. [You may want to translate into our symbolic language to assist your efforts.]

a. The Morning Star <u>is</u> the Evening Star.
b. John <u>is</u> a friend of Jill's.
c. Sloop <u>is</u> both judge and jury in trial.
d. The author of "The Sound and the Fury" <u>is</u> the same as the author of "A Rose for Emily."
e. No one <u>is</u> exempt from our club's rules.
f. Jill <u>is</u> the mother of John.
g. There <u>are</u> hard-working students.
h. Either there <u>will be</u> a sea-battle tomorrow or not.
i. Everything is material.
j. John is the best student.
k. The head of a horse is also the head of an animal.

(5) Translate the propositions expressed by the following meaningful sentences into PRED. To express uniqueness, you will need to choose from the following array of available symbols: functions symbols, definite description regimentation, and/or relational predicate symbols and the identity symbol as needed. Notice that the arity of a relational predicate is higher by one degree relative to the arity of the function symbol. For example, "the mother of John." Specify the key you are using.

$\boxed{m(j)}$ or $\boxed{\exists!xMxj \equiv \exists x\forall y(Myj \equiv y = x)}$

a. The mother of John is the sister of the father of Mary.
b. If the mother of Mary is married to the grandfather of John, then Mary is not married to John.
c. The grandmother of Mary is married to John only if the father of John is not married to the mother of Mary.
d. The capital of Secundia is to the north of the capital of Macabria.
e. A person's maternal grandmother is not that person's paternal grandmother.
f. The tallest student is John. [If we use a function symbol, the input of the function has to be a symbol for John's class.]

g. John's mother is Mary only if Mary's only son is John.

(6) Translate the following into PRED. Produce disambiguations, as needed.

a. None of the Fs are R-related to anything but Gs.
b. The Gs that are S-related to the Fs are R-related to each other.
c. Certain Fs are R-related only to themselves.
d. All Fs that are R-related to themselves now are R-related to each other always. [Use a constant, for instance "t_0" to symbolize the indexically specified present moment.]
e. If a Fs, then someone will G. [Unrestricted domain; a is person; time-periods need to be distinguished.]
f. If a is ever related to anyone, it must be to b.
g. If someone Fs, then someone Gs. [Disambiguations needed.]

(7) Translate the propositions expressed by the following idiomatic sentences of English into PRED+{$\perp$, $\not\equiv$}.
[Should we take idiomaticity to be a matter for the formal study of logic or something that falls under what we called Pragmatics?]

a. If Mary is not honest, then nobody is. [But you need to consider: is the consequent presented as an absurdity? Or is the point to reinforce that Mary is indeed honest.]
b. If Mary is the president of the Speluncean Club, then nothing makes sense anymore.
c. An honest politician is like a car without wheels. [Is the point that there are no honest politicians? Can we translate by means of an implication, to stay closer to the surface grammar of the sentence? Does the consequent state an absurdity? Does it matter that this is not logical absurdity but some other kind of impossibility - perhaps we should call it "technical implausibility"?]
d. And they lived happily for ever and ever after they married. [You will have to give the persons labels or names. It is implied that two individuals are involved.

How do we capture the expression "for ever and ever"? The point here is not to state an absurdity.]

e. If Mary does not understand this proposition, then this proposition doesn't make sense. [You will have to use a predicate for "proposition" and express uniqueness for "this proposition." The predicate for understanding will have to be binary (x understand y). Consider what we mean by "a proposition is equivalent with logical absurdity."]
f. It is absurd to state that there are exactly n prime numbers. [Could we get away with simply negating the claim that there are exactly n prime numbers? Can we express this proposition in PRED? How? Should we take it that "there are exactly n prime numbers if and only if logical absurdity" or "if there are exactly n prime numbers, then logical absurdity?"]

III A NATURAL DEDUCTION SYSTEM FOR PREDICATE LOGIC: PREDND

It is possible to build a natural deduction system for predicate logic. We do this by extending PROPnd*, the proof-theoretic formal system we constructed for the classical Propositional Logic, by adding four rules. The extension yields the system we will designate as PREDnd. The rules are needed for managing the introduction and elimination of the new symbols which are introduced grammatically for the universal and existential quantifier. In the proof-theoretic approach, the meanings of the connectives of a formal system are completely and exhaustively specified through the formal rules of the system. Linguistic justifications can be adduced, although they would have to be understood as external to the transactions of the formal system itself and not influencing it in any way.

The rules for the quantifiers constitute a rather complicated affair, which is not expected on the basis of our experience with the system for the unextended Propositional Logic (the logic of unanalyzed propositions, as we might say.) Tweaking of rules is needed for the construction of a natural deduction system for predicate logic. Since we are interested in having a system that is incapable of proving theses that should not be provable - given the character of the standard predicate logic - then we are compelled to resort to imposing certain restrictions on the rules in addition to giving basic justifications for those rules. The quantifiers of predicate logic can be thought of as connectives if, and only if, the domain is finite. In that case, we could make the case that our extension of the proof-theoretic systems for the standard Propositional Logic requires adding introduction and elimination rules for two more connectives. In that case, the rules we need for the quantifiers will require analysis that may be reduced to references to the justifications for the rules for conjunction (semantically aligned with the universal quantifier for finite domains, in the semantic approach) or inclusive disjunction (which can represent the existential quantifier semantically for finite-domain modeling.) Occasionally, under the proviso of finite domains, we may appeal to our ready intuitions by adducing such justifications. Nevertheless, in the proof-theoretic approach, we do not make modeling or semantic-theoretic considerations available to us and we have no grounds for entering specification as to the cardinal size of the domain.

We will need two types of schematic variables for constants in presenting the rules: we will use superscripts added to the constants for certain purposes. This unusual notational variation does not generate an alternative or non-classical logic: if we were to dedicate separate variable symbols to two domains, for instance, that would amount to constructing a formalism for an alternative logic. On the other hand, the standard predicate logic textbooks impose restrictions on the kinds of constants that are to be introduced in the course of a proof, depending on what quantifier is manipulated symbolically, but this task is relegated to the metalanguage and relevant indications are furnished in the justification-lines of the proof. Instead, we venture to formalize superscripts for the constant letters we use in the proof lines themselves. We import the grammar of PROPnd*. The superscript markers are originated from considerations we will present. The two kinds of superscripts correspond to different designations of constants: one we call arbitrary constant and use for subscript symbol "α" and the other we call new or non-arbitrary constant and we use as subscript the symbol "η". We will explain in due course what this is all about as we impart restrictions. Although we are working with an extension of PROPnd*, which yields PREDnd, we will required four rules for the quantifiers, which are introduction and elimination rules. PROPnd* dispenses with the introduction-elimination nomenclature but our quantifier rules will be presented as introduction-elimination rules. For the sake of convenience, we will also include a rule for quantifier-conversions: although this type of rule can be derived from the others, we aim at making proofs easier by accommodating an addition rule for conversion, which we symbolize by "∀∃". Moreover, since we will construct our formal system as all-encompassing, we will need inferential rules for identity between constants. We have seen before that a rule schema may encompass more than sub-schemata. It might be better, from a purist viewpoint, to think of the sub-schemata as being each a rule on its account.

There are important theoretical considerations for proceeding by steps, starting with Monadic Predicate Logic, which only has monadic predicate letters, and then proceeding to the introduction of n-place predicate letters and from there to the introduction of identity and even of functions. Metalogical properties are not shared between Monadic and Polyadic Predicate Logic systems. This is a deep theoretical result. Although we will point out some of the issues involved, we take the liberty in this text of presenting PREDnd as the comprehensive system that proof-theoretically represents the case of Polyadic Predicate or First-Order Logic with Identity.

We sanction elimination of the universal quantifier in a linguistic context when, as competent users of the language, we allow valid inference to a quantification-free statement from a universally quantified statement. Given the meaning of the universal quantifier, we should accept as valid an inference to a statement about any one of the objects. If the premise states that a property is possessed by all objects, then it validly follows that this property is possessed by any object whatsoever; thus – or, a fortiori, we might say – the property is possessed by any random or arbitrary object we may think of. If we glance sideways at the model-theoretic approach we presented in previous section, we find this rationale to be straightforward, and indeed to depend on the justification we give for elimination of conjunction, insofar as we limit our discourse to finite domains. The statement about all objects having an F-property is, by definition, the same as the conjunctive statement that the first labeled object has the property, and the second does, and so on. The objects are all presumed to be labeled by names which are simply tagging devices. Thus, by eliminating the conjunction we can generate a valid inference to the statement about any one of the objects possessing the F-property. In a natural deduction system like PROPnd* the rule we are putting to use has been called Simplification. If we reason along such lines, we can actually treat quantifiers like connectives and we can think of the quantifier rules as connectives rules that are added to yield PREDnd. The interesting question arises for infinite domains. The familiar conjunction we have been working with is not defined over an infinite number of conjuncts – at least, we have done nothing to attempt to so define it. The reasoning we have to fall upon to justify elimination of the universal quantifier is not straightforward in that case but it can go through. We reason that, if it is stated that all items have an F-property, then any one of the items must have this property. The meanings of words like "any" and "all" account for this. These are logical notions: so, we have done nothing that can be accused as smuggling extra-logical notions into our rationale. Our universe objects can still be considered as capable of being named, in the sense of being labeled: even though we can never effectuate this labeling process empirically, the members of such sets as the natural numbers or the rational numbers, and so on, are adequately available for our labeling purposes.

Now we present the rule for the Elimination of the Universal Quantifier, ∀E. The well-formed formula symbolized in the inferential schema by "ξ" is presumed to contain the individual variable of the quantifier as free – so that this variable is bound effectively by the quantifier. Since we are not restricted to Monadic Predicate Logic, we are already confronted with the task of accommodating elimination rules for iterated quantifiers. We use special symbols for the schemata and we start by specifying what these symbols are. It is to be understood that we retain the symbolic notation for PRED we have already introduced. We need different symbols, however, for the representation of the schemata. We make one exception for the symbol for identity in the schematic notation, which coincides with the familiar identity symbol of the object language. As with the object language, superscripts can be used to indicate arity (number of variable places following the predicate letter), although such may be omitted without risk of ambiguity. For the special superscripts we have dictated for the object language, regarding the two types of variables, we retain the symbols for the schematic notation.

NOTATION for INFERENTIAL SCHEMATA
Individual Variables: u, v, ..., u_j, ...
Individual Constants: λ, μ, ..., λ_i, ...
Predicate (Non-Logical) Constants: Φ, Ψ, ..., Φ_i, ...,
Well-Formed Formula: ξ
Universal Quantifier: Π
Existential Quantifier: Σ
Superscript Markers: α, η

III.1 ELIMINATION RULE FOR THE UNIVERSAL QUANTIFIER

⋮

k. $\Pi u \xi$

⋮

n. $\xi(\ldots \lambda^{\alpha}/u \ldots)$ ΠE(k)

Let us read the instruction embedded in the rule schema above. Given a universal quantifier attached to the well-formed formula (wff), such that the universal quantifier binds the associated free variables in the wff, the elimination rule compels removal of the universal quantifier symbol and uniform and complete substitution of the same *arbitrary* constant for all the individual variables that were bound by the universal quantifier. The substitution is uniform for all occurrences of the individual variable bound by the universal quantifier: all of them must be replaced by some, the same, individual constant with the "α" superscript to mark the arbitrary character of the constant – meaning that this could be any constant. Arbitrariness means that any individual constant can be used, unless there is some other restriction – to be specified – that applies. These replacement-operations are syntactic but can be defended by appealing to semantical considerations. Thus, we may reason that we have a valid inference from "everyone loves himself or herself" to "Jack loves himself," which requires a replacement of the pronoun "himself" uniformly by the name of Jack, but the inference is invalid to "Jack loves Jill" which is generated by non-uniform replacement of the sentence's pronoun by two different names. Hence the emphasis on uniformity of replacement.

Now we specify the specific conditions that apply to the arbitrary type of superscript. This superscript licenses substitution of any individual constant symbol whatsoever – whether it is a constant that has already appeared in a proof or a new constant. But we may have to impose some restrictions in this case too. We start with the restriction we have discussed already.

⇨ Restriction 1: Quantifier symbols are eliminated from left to right. They are introduced from right to left.

⇨ Restriction 2: In the case of consecutive iterated variables bound by the same universal quantifier symbol in the formula we replace the variables by means of the same individual constant. This can be any arbitrary constant. It can be a new constant too – except for when restriction 9, studied below, applies. An obvious corollary is that universal quantification by means of one quantifier *over two iterated variables* of a dyadic predicate is not permitted to yield two different constant symbols: the same constant symbol is to be iterated twice or – for nadic predicate symbols generally – n times. The grammatical arrangements that have been made for our formal language also take care of this, because having two different variables bound by one quantifier would mean that one of the variables is free: although this is still well-formed grammatically, the semantical or modeling characterization we have imposed tells us that this formula would not be expressing a proposition. It should be noticed that, for nested quantifiers binding associated variables, the same or different instantiating letters can be used for eliminating two different universal quantifiers; so, there is no restriction that different constant letters are to be used for the eliminations of different universal quantifiers.

The restriction we applied, and the restriction not needed, can be shown in examples like these:

1. ∀xFxx
2. Fab ∀E(1) WRONG! [violation of restriction 2]

1. $\forall x \forall y(Fx \supset Gy)$
2. $\forall y(Fa^{\alpha} \supset Gy)$ ∀E(1)
3. $Fa^{\alpha} \supset Gb^{\alpha}$ ∀E(2) PERMITTED>
4. $Fa^{\alpha} \supset Ga^{\alpha}$ ∀E(3) PERMITTED>

The second restriction can be readily justified. Consider what is wrong with the first proof shown above. Failing to observe it would allow us to infer from "everyone loves herself" that "Jill loves John!" On the other hand, for nested quantifiers, universal quantification is obviously permissive. If it is true that "everyone loves everyone" then it should follow both that "John loves Jill" and "Jill loves John" – and that "John loves himself" and "Jill loves herself" for that matter. Except for the restriction noted, universal elimination is one of the permissive rules in predicate logic formalisms. If we draw on our semantic intuitions, we may stipulate restriction to finite domains for the sake of providing a straightforward justification: if we think back to the rule simplification – or conjunction elimination – of natural deduction systems for Propositional Logic, we notice a similar permissive nature to the rule when it comes to what can be deduced. Every one of the conjuncts can be deduced by this rule. In the case of a finite domain for predicate logic, the conjuncts are propositions that state that the property in question is possessed by every one of the domain's objects. Accordingly, we may infer, by conjunction elimination, to any one of those propositions which yields an instantiation of the property for the corresponding object. (Universal Quantifier Elimination is sometimes called in textbook Universal Instantiation.)

We stipulate that the superscripts may be dropped in a subsequent line, indicated in the justification-line by "α-". We further liberalize the rule by permitting that removals of the superscripts may be effectuated "spontaneously" without need to justify this move and that this can be done at any point insofar as there is no risk of confusion. It is important to keep track, if not visibly then by some metalinguistic device, of the characterizations of constant variables and the attendant regulatory restrictions. Failure to apply or observe such restrictions would sanction proofs of sequents that are not valid in the standard predicate logic. Hence, the tweaking is obligatory and cannot be dispensed with.

Instances of application in PREDnd:

1. $\forall yFy$ P
2. Fa^{α} $\forall E(1)$
3. Fa η-(3)

1. $\forall xRxx$ P
2. $Rb^{\alpha}b^{\alpha}$ $\forall E(1)$

1. $\forall xRxx$ P
2. $Rb^{\alpha}b^{\alpha}$ $\forall E(1)$
3. Rbc η-(2) WRONG! [violation of restriction 1]

1. $\forall x(Fx \equiv Gx)$ P /
2. Fc P /.. $Fc \supset Gc$
3. $Fc^{\alpha} \equiv Gc^{\alpha}$ $\forall E(1)$
4. $Fc \equiv Gc$ η-(4)
5. $(Fc \supset Gc) \cdot (Gc \supset Fc)$ Equivalence(4)
6. $Fc \supset Gc$ S(5)

In general for iterated and nested quantifiers that are mixed (universal and existential), the order that is to be postulated by means of a subsequent restriction is *from left to right*. Witness the first proof below.

1. $\forall x \forall y Ryx$ P /.. Rba
2. $\forall y Rya^{\alpha}$ $\forall$E(1)
3. $Rb^{\alpha}a^{\alpha}$ $\forall$E(2)
4. Rba α-(3)

We can also prove the following. There is no need to impose constraints that block the inference in the following example. Semantic justification was given in the preceding paragraph. We may reason as follows. If it is given as a premise that everything is related to everything, this should allow for the inference to the effect that everything is also related to itself. Our rules allow us to derive this inference because the arbitrariness superscript licenses choice of any individual constants.

1. $\forall x \forall y Ryx$ P /.. Raa
2. $\forall y Rya^{\alpha}$ $\forall$E(1)
3. $Ra^{\alpha}a^{\alpha}$ $\forall$E(2)
4. Raa α-(3)

Given restriction 1, the following is not, as it should not be, derivable.

1. $\forall x Rxx$ P /.. Rab
2. $Ra^{\alpha}b^{\alpha}$ WRONG! [violation of restriction 1]

⇨ Restriction 3: Vacuous Quantification: We enforce a restriction that also made sense when we contemplated and structured the syntactical or grammatical setup for a predicate logic system. We ban vacuous quantification in a broad sense, defined as: a) having a quantifier symbol that has no variable to bind, or b) as a (vacuously) iterated quantifier attempting to re-bind a variable that is bound by another quantifier of the same kind (universal or existential.) Given this exception, the following proof is blocked, as it should be.

Semantically, we can reason that vacuous quantification is like a redundant added universal quantifier, something along the lines, "everyone and everyone else love themselves" or, for the second clause of the definition of vacuous quantification, having a statement like "everyone Jill loves herself," which is clearly a solecism (meaning a wrong use of language.)

1. ∀x∀yRxx	P	/.. Rab	(1) [Restriction 3] [We cannot proceed.]

Lacking Restriction 3 would generate the following terminating proof procedure, which clearly allows us to prove an invalid sequent. Note what happens: in line 2, we are compelled to match the quantifier variable to the individual variable in the formula within the quantifier scope.

1. ∀x∀yRxx	P	/.. Rab	
2. ∀yRa$^{\alpha}$y	∀E(1)		
3. Ra$^{\alpha}$b$^{\alpha}$	∀E(2)		
4. Rab	η-(3)	!	

⇨ <u>Restriction 4:</u> Main-Connective Restriction. Regardless as to whether we treat quantifiers as connectives or not – based on considerations about the cardinality of the domain, if it is finite or not – we impose a restriction that is worth institutionalizing. We set it as a restriction, which we call the Main-Connective Restriction. This prohibits making replacements of variables bound by quantifiers when the formula has a main connective other than the quantifier (assuming that we are thinking of the quantifier as a connective.)

Semantically, we can consider as an example, an inference from "If everyone is in love, then everyone is happy" to "If Jill is in love, then John is happy." This should not be accepted as a valid inference but it would go through as such if we permitted eliminations of the universal quantifiers regardless of the fact that the main connective in this statement is material implication. The truth of the statement that Jill is in love should not allow deduction of a statement to the effect that John is happy; what we are given is that, if everyone is in love then everyone is happy. The statement asserting as fact that Jill is in love is not equivalent to the statement asserting that everyone is in love.

Let us assume, for the sake of illustration and considering finite domains, that John and Jill are the only two members of a restricted domain. The statement "if everyone is in love, then everyone is happy" means, expanded by treating universal quantification as a conjunction, that "if John and Jill are in love, then John and Jill are happy." The purported conclusion, "if Jill is in love, then John is happy" does not follow validly from the given premise, as we can see by applying familiar standard Propositional Logic.

$((La \cdot Lb) \supset (Ha \cdot Hb)) \nvDash Lb \supset Ha$

A counterexample, or countermodel, can be easily generated.

$\surd(a) = \text{John}$
$\surd(b) = \text{Jill}$
$\surd(L) = \{b\}$
$\surd(H) = \{b\}$

Thus we have:

$((La \cdot Lb) \supset (Ha \cdot Hb)) \nvDash Lb \supset Ha$
F F T T F F T T F F

1. $\forall xLx \supset \forall xHx$ P /.. $La \supset Hb$
2. $La^{\alpha} \supset \forall xHx$ $\forall E(1)$ [violation of restriction 4]
3. $La^{\alpha} \supset Lb^{\alpha}$ $\forall E(2)$ [violation of restriction 4]

III.2 ELIMINATION RULE FOR THE EXISTENTIAL QUANTIFIER

⋮
k. $\Sigma u\xi$
⋮
n. $\xi(\ldots \lambda^{\eta}/u \ldots)$ $\Sigma E(k)$

This rule is expected to carry significant weight in the calibration of proof-theoretic systems for predicate logic. It can be readily justified. The key to what is being done is that elimination of the existential quantifier (in some textbooks called Existential Instantiation) requires introduction of a constant letter that is brand new - not hitherto used in the proof. We call this constant "new" or "non-arbitrary" and it is to be thought of as a constant that should not have occurred in the proof already; in a broad sense, it cannot be any constant whatsoever - hence, it cannot be an arbitrary constant. To understand how this is justified, let us reflect on the statement "someone is in love". We have no sanction to infer validly that a specific person is in love. It is irrelevant whether we happen to know of such a fact - predicate logic, like deductive logic in general, is not moored in empirically ascertainable states of affairs or events. We must take pains, then, to introduce a name - a constant letter - that is random and has no special connection to any other name. Most emphatically, this has to be a name that may or may not be possessing the property but about which we are making a provisionary assumption that it possesses the property. In this way, we effectually insulate the object we are talking about from any other proof-theoretic considerations. We may not presume to know anything else about this object. Therefore, we have to ensure that this object does not happen to have any other specific characteristics whatsoever. If anything is already known about this object, then the object is privileged in a sense - it is not new - and we have no warrant for making the assumption that this particular object is an exemplar for the property presented in the existential-quantification statement. This is what we mean that the name or constant that is to be used in the theoretical construct is new in the flow of the proof.

Of course, a subsequent universal elimination may be applied to the object that has been named by the new constant. If a statement is universal, then it applies to all objects and, a fortiori, it applies to our new object whose mention was induced by existential elimination. There is, however, a catch: Crucially, a universal elimination that is applied before the existential elimination, using a constant letter, effectively takes this name out of consideration when it comes to a subsequent existential elimination. We can think why this is so. If we infer, from "everyone is in love" to "Jill is in love", then a statement, taken up subsequently, to the effect that "Someone is happy" does not at all sanction inference to "Jill is happy." It could be someone else – this is obvious. This illustrates neatly why we absolutely need a new name. This case also highlights the difficulties that arise from working with mixed quantifiers in a formula and also the role played by the order in which rules for the quantifiers are applied. We will have more to say on this subsequently. We are ready now to state our next restriction. Depending on what proof-theoretical arrangements are made, it is possible to have fewer or more restrictions but this one restriction is ubiquitous.

- ⇨ Restriction 5: New Constant for ∃E. Elimination of the Existential Quantifier compels introduction of a new constant variable uniformly replacing the individual variable bound by each existential quantifier that is being eliminated. The superscript "η" indicates that the constant is new, and, as such, not appearing in the proof so far.

Failure to observe this restriction terminates pseudo-proofs that incorrectly derive conclusions that cannot be validly derived from the given premises in predicate logic.

III.3 INTRODUCTION RULE FOR THE UNIVERSAL QUANTIFIER

⋮

k. $\xi(\ldots \lambda^{\alpha} \ldots)$

⋮

n. $\Pi u \xi (\ldots u/\lambda^{\alpha} \ldots)$ $\Pi I(k)$

The semantic justification for this rule is straightforward. Universal quantification can be applied only in cases in which claims have been made strictly about an arbitrary object, which could be any object whatsoever. It is crucial that we specify what this means exactly. It does not mean that given a name – constant – we can treat this as an arbitrary case. Any given name has to be regarded as non-arbitrary. Basically, it is then regarded as what we have called a "new" constant in the proof and it is to be treated as such. A safe way to grasp what arbitrariness means in this context is by considering that we have, theoretically, ascertained that every single one of the objects in the domain has a certain property; hence, we can infer validly that every object has the property. We should think, then, of a conjunction of propositions – each one of which, propositions, states that each one of the objects in the domain has the property in question. In the proof-theoretic systems for Propositional Logic, this corresponds to the rule of conjunction or conjunction introduction and we can appeal to this assuming that we are working with finite domains. In informal proofs in Mathematics and in other contexts, we may stipulate some arbitrary object and name it but, for our purposes, given constants are treated as "new", which means for our case that they are non-arbitrary but specific.

Matters become more complicated when mixed quantifiers are in the formula. By mixed quantifiers we mean that both universal and existential quantifiers are in the well-formed formula.

III.4 INTRODUCTION RULE FOR THE EXISTENTIAL QUANTIFIER

⋮

k. $\xi(\ldots \lambda^{\eta} \ldots)$

⋮

n. $\Sigma u \xi(\ldots u/\lambda^{\eta} \ldots)$ $\Sigma I(k)$

⋮

l. $\xi(\ldots \lambda^{\alpha} \ldots)$

⋮

m. $\Sigma u \xi(\ldots u/\lambda^{\alpha} \ldots)$ $\Sigma I(k)$

A convenient fact about the existential quantifier is that it can be introduced validly on the basis of any constant letter whatsoever – regardless of how the letter was generated in the first place (thus, regardless as to whether it is an arbitrary or new constant). Semantically justifying this, we can say that even if it is true that "Jill is a student" is derived validly from "Everyone is a student", it must still be true that someone is a student insofar as the person named Jill is a student. The standard predicate logic can be challenged on philosophic grounds for this profligacy but it is a fact that, as our rule for the elimination of the universal quantifier shows, we have exemplification for any statement in which the main connective is the universal quantifier. This means that "all things have the F property" validly implies "a has the F property", where ⌜ a ⌝ is an arbitrary constant that labels some distinct object in the domain. The formalism is set up in this fashion and, when we have introduced the identity sign and rules for it, we will see that the following statement is a tautology in the standard predicate logic: "for everything x, there is at least one thing y, such that x and y are identical." The formal fact that this statement is a logical truth seems mysterious at first but we can parse it like this, to see that it articulates a certain kind of existential permissiveness. "Anything presumed to be in our model's domain has at least one name that denotes it." Because denotation suffices to establish existence in the domain – we track only existing things by labeling – we are essentially saying that everything that is in the model-domain cannot fail to exist. This statement is surely trivial or tautologous. Of course, this all means that existence is relativized to domains of models. This might seem unintuitive, since existential statements are synthetic rather than granted meanings by definitions of logical terms. On the other hand, it can be defended on the grounds that, precisely because existence is not a logical matter, our formalized languages should be applicable only in cases for which matters of existence are already considered as having been settled – for instance, in the case of fragments of language that are used by science. Refusing to take a metaphysical position requires, on this view, relativizing existence to domains of models. This is a deep philosophic subject and there are alternative, non-standard approaches, which we do not present in this text, allowing for alternative options as to how to treat existential commitments.

We consider an example. We give extensive justifications in the margin (metalinguistically) and we refer to the lessons of the natural deduction system we constructed for the basic Propositional Logic – which is, after all, extended, with added rules and restrictions, to yield the present Predicate Logic system.

1. $Fa^{\eta} \supset \forall xGx$ P /.. $\exists x\forall y(Fx \supset Gy)$	[Any given constant is non-arbitrary]
2. Fa^{η} AP	[Assumed Premise, for Conditional Proof]
3. $\forall xGx$	[Modus Ponens (1, 2)]
4. Gb^{α}	[$\forall$E(3) – arbitrary constant]
5. $Fa^{\eta} \supset Gb^{\alpha}$	[Conditional Proof (2-4) – discharging AP]
6. $\forall y(Fa^{\eta} \supset Gy)$	[$\forall$I(5) – note the introduction from right to left]
7. $\exists x\forall y(Fx \supset Gy)$	[$\exists$I(6)]

We can also prove the following, which is also a valid sequent. Note how we make different moves in the proof-process.

1. $Fa^{\eta} \supset \forall xGx$ P /.. $Fa \supset Ga$
2. Fa^{η} AP [Assumed Premise, for Conditional Proof]
3. $\forall xGx$ [Modus Ponens (1, 2)]
4. Ga^{α} [$\forall$E(3) – arbitrary constant which can be the one in use]
5. $Fa^{\eta} \supset Ga^{\alpha}$ [Conditional Proof (2-4) – discharging AP]
6. $Fa \supset Ga$ [Dropping superscripts from 5]

⇨ Restriction 6: For consecutively iterated or nested existential quantifier symbols that are introduced, different variable symbols must be attached to the introduced existential quantifiers.

This makes eminent sense semantically. Given that Jack loves Jill, it does not follow necessarily or validly that Jack loves himself. If, however, we allow for the same pronoun to be used, we derive from "Jack loves Jill" that "there is someone who loves himself or herself." This captures the semantic flavor of what is wrong with allowing iterated identical variable symbols. This gives us an idea as to what is wrong with repeating the same variable letter. Failure to observe this restriction in the following fallacious proofs allows us to derive conclusions that are not validly derivable in predicate logic.

1. Rab — P /.. $\exists yFyy$
2. $\exists yFay$ — $\exists$I(1)
3. $\exists y\exists yFyy$ — $\exists$I(2) WRONG! [violation of Restriction 6]
4. $\exists yFyy$ — [given Restriction 3 on vacuous quantification]

1. $\exists xRax$ — P /.. $\exists xRxx$
2. $\exists x\exists xRxx$ — $\exists$I(1) WRONG! [violation of Restriction 5]
3. $\exists xRxx$ — [given Restriction 3 on vacuous quantification]

⇨ Restriction 7: New Constant Trumps Arbitrary Constant. If the same constant variable is available in the proof both as arbitrary and new, it is to be considered as new for purposes of making further moves in the proof process. This dual availability of the same constant letter can easily happen. For instance, existential quantifier elimination could yield a new constant, λ, and that same constant can subsequently be obtained by applying universal quantifier elimination. This is permissible since the new constant is certainly one of the names labeling some object in the domain; by eliminating the universal quantifier, we may use any constant or name – this is what "arbitrary" means in this context – and, so, we can use λ. Nevertheless, the character of the constant λ, so to speak, is stamped by the mode of its generation, which makes it a new constant. This means, for instance, that we cannot appeal to the presence of λ to introduce a universal quantifier: even though λ has also been generated by universal elimination, it remains "genetically" the case that its character is that of a new constant – because "new trumps arbitrary" – and it cannot be appealed to for introduction of universal quantification.

Semantically speaking, let us consider the following example. Every student takes classes – which we read in our formalism as "for everything, if it is a student, then it takes classes." Jill is student. The name "Jill" corresponds in a linguistic context to what we designate as a new constant. We can infer validly, by elimination of the universal quantifier, that if Jill is a student, then she takes classes. And we have the premise according to which Jill is a student. Therefore, it is a valid inference to make that "Jill takes classes." If we could generalize, by treating "Jill" as an arbitrary name, we would derive "everyone takes classes", which is clearly an invalid inference. We can, however, validly infer that "at least one person take classes" because introduction of the existential quantifier is sanctioned on the basis of a new constant (like the name "Jill" in our example.) (We have been assuming throughout restricted domain to persons.)

Failure to observe Restriction 7 terminates fallacious proofs like the following.

1. ∀x(Fx ∨ Gx)	P
2. ∃x ~ Gx	P /.. ∀xFx
3. ~ Ga^{η}	∃E(2)
4. Fa^{α} ∨ Ga^{α}	∀E(1)
5. Fa^{α}	Disjunctive Syllogism (4, 5), but WRONG superscript!
6. ∀xFx	∀I(5)

1. ∀x(Fx ⊃ Gx)	P
2. ∃xFx	P/.. ∀xGx
3. Fa^{η}	∃E(2)
4. Fa^{α} ⊃ Ga^{α}	∀E(1)
5. Ga	Modus Ponens (3, 4) – but WRONG superscript! [violation of Restriction 6]
6. ∀xGx	∀I(5)

Obviously, we could tweak the formal mechanisms of the proof by legislating as to what quantifier may be introduced under given conditions like those in the preceding proof. We could impose a restriction on introducing the universal quantifier, which would be violated at line 6. We have the same result. It should be realized that these restrictions affect the justificatory machinery of the proofs and, as such, they are presented metalinguistically. The unusual license we have taken in letting the symbols take superscripts should not disguise that fact. Indeed, we can ascertain that the formal objects are not affected in a deeper sense: for example, witness how modus ponens is still applicable to yield line 5 regardless of the different designations of superscripts. We can account for this. The object denoted by the logical constant ⌜a⌝ is certainly the same regardless of what inferential license, or lack of inferential license, we have in the application of the proof-theoretic machinery.

⇨ Restriction 8: Given Constant as New. We impose a restriction to the effect that any given constant in a formula is to be treated not as arbitrary but as new. Indeed, we should not be able to infer validly from "Jill is a student" to "everyone is a student;" if we treated the name "Jill" is arbitrary, however, nothing could prevent an inference to the universally quantified statement given the rules we have presented.

Failure to observe this restriction is in evidence in the following fallacious proof.

1. ∀x(Fx ⊃ Gx) P
2. Fa P/.. ∀xGx
3. Fa^{α} Reiteration(2) WRONG! [violation of Restriction 7]
4. $Fa^{\alpha} \supset Ga^{\alpha}$ ∀E(1)
5. Ga^{α} Modus Ponens (3, 4) [the superscript is not wrong; the error is in line 3]
6. ∀xGx ∀I(5)

⇨ Restriction 9 – Prenex-Permission: We stipulate that no rule implementation can result in a prenex form that is wrong relative to the arrangements discussed in I.1.2. This restriction is motivated by purely pragmatic considerations. The transformations into prenex forms we discussed in I.1.2 ought to be provable by the use of the rules in the present system. By adding this restriction, we add redundancy and we essentially introduce a permission: we may, although we are not obligated to, transform formulas into their equivalent prenex forms before we continue to apply natural-deduction rules in a proof.

We consider the following example of an invalid empty-premise sequent that should not be provable within a proof-theoretic system that harmonizes properly with the semantics of the standard predicate logic. Semantically approached, this proof would be tantamount to sanctioning as a logical truth the claim that "everything is such that if it Fs, then everything Fs." There may be properties that are such that this makes sense but it should not be a truth of logic since there are surely properties for this is not the case. Of course, we cannot undertake a full transparent expression of the justification just given in our symbolic notation because that would require resources of a second-order logical system – given that we want to say "there is some property, such that___" which requires quantification over predicate symbols.

$\nvDash \forall x(Fx \supset \forall xFx)$

1. $Fa^{\alpha} \supset Fa^{\alpha}$ Thesis [Introduction of Thesis of PROPnd]
2. $\forall y(Fa^{\alpha} \supset Fy)$ ∀I(1) ERROR! [uniformity of bound variables violated]
3. $\forall x\forall y(Fx \supset Fy)$ ∀I(2)
4. $\exists xFx \supset \forall yFy$ Prenex(3) [reversing prenex transformation]

We have reached a formula that is not a thesis of Predicate Logic although we initiated the proof procedure with a thesis! This is due to the error at moving from 1 to 2. The prenex transformation detects the error.

In contrast, the following is derivable. Semantically approached, this means that "there is at least one thing such that if it Fs, then everything Fs." Even though this is not immediate to detect, the claim is rather weak – it is about something, at least one thing – and, since the one-member domain is not logically impossible – it is not to be excluded presumptively from consideration – this should go through: it is logically possible that something having an F-property implies that everything there is in the domain has this property.

$\vDash \exists x(Fx \supset \forall yFy)$

1. $\forall xFx \supset \forall xFx$ Thesis of PROPnd

2. ∀xFx	AP
3. ∀xFx	MP(1, 2)
4. Fa^{α}	∀E(3)
5. ∀yFy	∀I(4)
6. ∀xFx ⊃ ∀yFy	CP(2-5) [discharging of AP]
7. ∃x(Fx ⊃ ∀yFy)	Prenex(6)

⇨ Restriction 10: When existential elimination has been used after universal elimination, in reversing the order to apply introductions, universal quantifier introduction is not permitted. In the formal schematization we have been constructing, this can also be represented as making a permission for introduction of the universal quantifier only in the case in which we can have same constant letters (regardless of the varying superscript types of those constant letters.) [This latter permission overrides, only for this case, the earlier restriction by which new constants trump arbitrary constants.]

Once again, we see that, while in the case of universal quantifiers, the order is unimportant, this is not so for mixed quantifier symbols.

Justification for restrictions 7 and 8 can be offered now. Essentially, such justifications recapitulate what can be said to show that a certain quantifier-shift, which we discuss further below, is invalid. We cannot validly proceed from "everyone is R-related to someone" to "there is at least one specific someone to whom everyone is related." The converse, however, should be valid. The restrictions we have put in place ensure that the illicit shift cannot proceed. Depending on what type of proof-theoretic formalism is presented, textbooks have what appear to be different ways of dealing with this issue but the underlying issue is always the imperative of blocking the illicit quantifier switch. Often, textbooks speak of a problem of dependency and draw on this notion to enforce appropriate formal proof-theoretic restrictions. The dependency is of an attempted introduction of a constant for universal quantifier introduction on a constant that has been made available by means of existential elimination. Assuming that everyone loves someone permits us to provisionally introduce one name for the existential elimination and any name for the universal elimination; nevertheless, when we introduce quantifiers subsequently, starting with the new name that labels the person loved, we should not infer that anyone can now be talked about. The person loved has not been asserted to be loved by everyone and, therefore, references to her should not be used to sanction claims about any arbitrary person loving her. This is the gist of the matter but it is left to the formal schematic arrangements to manage proof-theoretical moves in such a way as to block such an invalid inference. In some proof systems, this potential problem can be handled without requiring re-adjustments of the formal machinery.

This is a good example of how, unexpectedly perhaps, predicate logic formalisms depend on motivated insertions of special arrangements – permissions and restrictions – which presents a situation that never arises in the unextended Propositional Logic.

- ⇨ Restriction 11: Let us define as saturation the proof-theoretic process by which elimination of the universal quantifier is applied as often as it takes so that all available constant symbols are used. We may want to avail ourselves of this device in a proof. There is a problem, however, in that predicate logic can run into non-terminating proof procedures – something that never arises as a specter over proof methods in Propositional Logic. Suppose that we have an unrestricted domain, for which a denumerable infinity of individual constant symbols (of the size of the positive integers) has been made available; circumstances arise in which introduction of yet another constant is available so that the proof procedure suffers structurally from a non-terminating prospect! We make a stipulation at this point, and impose a restriction which we will carry over along with the other restrictions when we come to the construction of a tree method for predicate logic. According to the Saturation Restriction, we may *add* constant symbols up to saturation (covering all the available constant symbols) when we understand as "available" only those constant symbols that have been introduced by existential elimination. This applies only when we go beyond the initial steps – note the use of the word "add" above. We certainly introduce constants by means of universal elimination in a proof but the point of this restriction is that we cannot continue after all quantifiers have been eliminated toward additional applications of universal elimination.

We present examples. The first example shows that iterated universal quantifiers can be re-lettered systematically for the individual variables by means of valid inferences from lines to succeeding lines. The second example shows how an invalid sequent is provable if restriction 6 or restriction 8 is violated. This invalid or faulty proof – we might call it a pseudo-proof – derives a conclusion that has an illicit shift in the order of the quantifiers. We will elaborate on this fallacy in a subsequent paragraph but we see, first, how violations of restrictions can result in deriving purported conclusions that do not follow validly from the premises. In the first faulty proof, the procedures of elimination begin from the left instead of from the right, thus violating Restriction 6. This defect permits use of the same constant letter when the universal quantifier is eliminated and the subsequent introductions of quantifiers are maneuvered to produce the result – which is an illicit in the quantifiers. In the second faulty proof, it is restriction 8 that is violated.

1. ∀x∀yRxy/.. ∀y∀xRyx
2. ∀yRay ∀E(1)
3. Raa ∀E(2)
4. ∀xRax ∀I(3)
5. ∀y∀xRyx ∀I(4)

<u>1st Incorrect Proof of the Invalid Quantifier Shift</u>

1. ∀x∃yRxy /.. ∃y∀xRxy
2. ∀xRxa^{η} ∃E(1) WRONG! [violation of restriction 1]
3. R$a^{\alpha}a^{\eta}$ ∀E(2)
4. ∀xRxa^{η} ∀I(3)
5. ∃y∀xRxy ∃I(4)

<u>2nd Incorrect Proof of the Invalid Quantifier Shift</u>

1. ∀x∃yRxy /.. ∃y∀xRxy
2. ∃yRa^{α}y ∃E(1)
3. R$a^{\alpha}b^{\eta}$ ∀E(2)
4. ∀xRxa^{η} ∀I(3) WRONG! [violation of restriction 10]
5. ∃y∀xRxy ∃I(4)

In this way we have proven an invalid sequent, known as the Birthday Fallacy. To consider an instantiation of the argument form, from which the popular name is derived, it is not valid to infer from "everyone has some birthday date" that "there is one birthday date which everyone shares." This is an illicit quantifier shift which is, alas, ubiquitous in illustrious philosophic deductions – like the one from "everything has a creator" to "there is one creator of everything" to which such luminaries as Aristotle and Aquinas appear to have succumbed. The quantifier shift from ∀∃ to ∃∀ is, accordingly, illicit but the shift from ∃∀ to ∀∃ is not as the following proof, which does not violate any restriction, shows. Indeed, if it is true that "there is someone everyone loves" then it must also be true that "everyone loves someone."

1. $\exists x \forall y Rxy$ /.. $\forall y \exists x Rxy$
2. $\forall y Ra^{\eta} y$ $\exists E(1)$
3. $Ra^{\eta} a^{\alpha}$ $\forall E(2)$
4. $\exists x Rxa^{\alpha}$ $\exists I(3)$
5. $\forall y \exists x Rxy$ $\forall I(4)$

The re-lettering is not problematic, as it is permitted, and for good reason. The order of eliminations and introductions is observed as dictated by restriction 5. If, on the other hand, we attempted a proof of the illicit quantifier-shift sequent and tried to proceed without committing any errors, we would find ourselves unable to proceed at some point in the proof.

1. $\forall x \exists y Rxy$ /.. $\exists y \forall x Rxy$
2. $\exists y Ra^{\alpha} y$ $\forall E(1)$
3. $Ra^{\alpha} b^{\eta}$ $\exists E(2)$
4. $\exists y Ray$ $\exists I(3)$ WRONG! [not the same constant letters!]

By having extended PROPnd*, we have simplified matters. We can rely on a redundant abundance of inferential rules – rather than on the rules from a parsimonious system with introduction and elimination rules only as needed. Still, the proof method we called Conditional Proof is available in PROPnd, and a fortiori in PREDnd, and this fact calls for certain comments.

The informal proof method commonly used in Mathematics allows us to work with posits of arbitrary entities – which are to be accepted as such insofar as no special assumption is made about them and nothing in the proof requires reliance on any such additional assumptions. We cannot do this in PREDnd.

- ⇨ Conditional Proof (CP) Restriction. Any posited formula (or assumed premise) has to contain what is considered a new constant.

Otherwise, the following fallacious proof of a presumed tautology (empty-premise set sequent) would have to be accepted.

1. Fa^{α} AP WRONG! [violation of CP Restriction]
2. ∀xFx ∀I(1)
3. Fa ⊃ ∀xFx CP(1-2)

III.5 ∀∃ INTERCHANGE (OR ∀∃ CONVERSION, OR ∀∃ EXCHANGE) RULES

We facilitate proofs by including a comprehensive derivation rule that allows interchanges between the two quantifier symbols and between combinations of negation symbols and quantifier symbols.

⋮
k. Πuξ
⋮
n. ~ Σu ~ ξ ΠΣ(k)
⋮
l. ~ Πu ξ
⋮
m. Σu ~ ξ ΠΣ(l)
⋮
n. ~ Σu ξ
⋮
o. Πu ~ ξ ΠΣ(n)
⋮
r. Σu ξ
⋮
s. ~ Πu ~ ξ ΠΣ(r)

We include these rules for convenient simplification of proofs. It should be possible to derive these rules but those proofs can be quite complicated given especially that we have imposed the restriction regarding the main connective.

In natural language, as used in ordinary linguistic settings, there is often ambiguity as to quantifier scope but translation into a formal idiom like ours compels disambiguation: if ambiguity is presumed to be retained, we should think of our formulas as "unreadable." Given the defined meanings of the quantifiers in our formal language, and assuming that the corresponding linguistic meanings are unambiguously present, we may scan the straightforward semantic justifications. If it is true that everyone has property F, then it must be false that someone, at least one person, does not have the property F. If it is false that everyone has F, then it must be true that at least one person does not have the property F. If it is false that at least one person – even one person – has property F, then it must be true that everyone does not have the property F. And if someone, at least one person, has F, then it cannot be true that everyone does not have F.

III.6 ELIMINATION RULE FOR IDENTITY

⋮

k. $\lambda_1 = \lambda_2$

⋮

n. $\xi(\ldots \lambda_1 \ldots)$

⋮

l. $\xi(\ldots \lambda_2 \ldots)$ =E (k, n)

To justify this rule on a semantic basis, let us think of a case in which, for whatever reason, two different names identify exactly one, unique, individual. Whatever properties are attributed to this individual under the one name must be attributed to the individual under the other name: this means that the propositions expressed by property-assigning sentences that use the one name in the sentence have the same meaning (truth value) with the propositions expressed by the same sentences when only the one name is replaced by the other name of the individual. This works because names are assigned functionally – there is a unique object assigned to each name, something that is not the case in natural language in which different individuals can be identified by the same name; this device fails in the contexts we characterized as referentially opaque. For instance, one may know that Rob is a carpenter, and it may also be true that Bob is another name for Rob, and yet it does not follow that it must be known that Bob is a carpenter. Thus, substitution of "Bob" for "Rob" does not necessarily preserve the truth value; this is because "it is known that__" generates a referentially opaque context, as the case is with all modal phrases (like "necessarily," "possibly," "it is obligatory that__," "it is forbidden that__," "it has always been the case that__," and other such phrases.

III.7 INTRODUCTION RULE FOR IDENTITY

⋮

k. $\lambda_1 = \lambda_2$ =I

⋮

The rule for the introduction of the identity symbol can be defended readily once we keep in mind that identity is a matter of co-reference in predicate logic. Given that the constant symbols co-refer, they ought to be inter-substitutable within any context that is generated by a non-opaque extensionalist context. Semantically, if "Sally" is another name by which Mary is called, then any property possessed by the person named Mary ought to be had by the person named Sally and vice versa. The formal requirement that we have in the predicate formalism – not to be found in natural language – is that the names, "Mary" and "Sally" in our example, each have exactly one unique referent. In general, the referent (truth value or object or set of objects) is the logical meaning of a proper notional expression of an extensionalist system; given that two expressions have the same logical meaning (which means, in the case of constant symbols, that they refer to the same object of the domain), then, following the proper grammatical conventions of the system, these expressions are inter-substitutable within any well-formed expression. In other words, logical meaning cannot chance – it is preserved – when components are replaced by components that have the same meaning as the ones that are replaced. The meaning of the whole propositional expression is its truth value: thus, preservation of meaning is preservation of truth value for the propositional formula that has co-referring constants inter-substituted. Another way of saying this is that cross- substitutions of logical equivalents can be carried out *salva veritatis* (without changing the truth value) into propositional formulas. This is indeed the definition of what we mean by Extensionalism in logic.

The introduction rule for the identity symbol allows us to enter into a proof a line that, semantically interpreted, expresses a logical truth. Surely, a name co-refers with itself. We ensure this by legislating in our grammar that every constant has exactly one object it refers to – the reference is fixed, within the signature of the model, by means of a function that assigns a unique referent to each constant symbol. Notice, then, that the truism "everything is identical with itself" is represented within the formalism of standard logic as "every name labels one unique object."

We allow negation of identity to be symbolized by "≠".

We have equipped our notational formal system with expressive resources for functions. Function symbols denote or refer to objects in the domain; as such, they are terms along with individual constant symbols and the individual variable symbols (which are systematically non-referring.) An exceptional characteristic of function symbols is that they are defined as having a unique referent; thus, they may be uniformly replaced by individual constants but, in each such case of replacement, the constant has to be considered unique and, as such, restricted from further use when other function symbols are likewise replaced. We introduce a special superscript for uniqueness for this purpose, symbolized by "u" and placed in superscript position. Accruing as unique-superscripted constant symbols, they are to be considered as what we have called "new" constant symbols; as such, they cannot allow for introduction of universal quantification. (We get the same result based on our earlier restriction about treating any constants in the proof as new; hence, this rule is, strictly speaking, redundant.) Moreover, constant symbols used for eliminating function symbols have to be considered as identical (which means, co-referring), given the uniqueness of the function output. A function symbol is to be understood, in semantic rendering, as standing in for the phrase "the such-and-such." In terms of symbolic resources, the same result can be accomplished by using predicate letters but only if a formula is conjoined to express uniqueness of the object that has the property expressed by the predicate letter.

Having liberalized the superscript rules, we grandfather the liberalization and allow that the superscript can be omitted – which does not mean that the uniqueness restriction itself can be ignored. For purposes of using distinct symbols in the schemata for rules, we use for functions symbols members from $\{\vartheta, \ldots, \vartheta_j, \ldots\}$.

III.8 FUNCTION SYMBOLS RULES

Function Elimination Rule

⋮

$\vartheta(\lambda)$

⋮

a^u fE

Function-Identity Rule

⋮

$\vartheta(\lambda) = \mu^u$

⋮

$\vartheta(\lambda) = \nu^u$
⋮
$\mu = \nu$ f=

General case (nary function symbol)
⋮
$\vartheta(\lambda_1, \ldots, \lambda_n) = \mu^u$
⋮
$\vartheta(\lambda_1, \ldots, \lambda_n) = \nu^u$
⋮
$\mu = \nu$ f=

Failure to observe the uniqueness rule about function symbols would permit termination of the following faulty proof or pseudo-proof.

1. ∃xRxf(a) P1
2. ∃xSxf(a) P2
3. Rcf(a) ∃E(2)
4. Sdf(a) ∃E(3)
5. Rct_1^u fE(3)
6. Sdt_2^u fE(4) WRONG!
[As it is not remedied by identity: $Rct_1^u = Sdt_2^u$]
7. $Rct_1^u \cdot Sdt_2^u$ ·I(5, 6)
8. $\exists y(Rct_1^u \cdot Sdy)$ ∃I(7)
9. $\exists x\exists y(Rct_1 \cdot Sxy)$ ∃I(8)
10. ∃z∃x∃y(Rcz · Sxy) ∃I(9)
11. ∃w∃z∃x∃y(Rwz · Sxy) ∃I(10)

This would allow us to prove the following, which is patently invalid: assuming that a person's father has a brother and a sister, it would presumably follow that four distinct individuals are related as brothers and sisters.

III.9 EXAMPLES

a

1. ∀x(Fx ⊃ ∀y(Gy ⊃ Rxy)) P1

2. ∀x(Hx ⊃ Gx)	P2	
3. Ft^{η}	P3 /.. ∃x∀y(Hy ⊃ Rxy)	
4. Ft^{α} ⊃ ∀y(Gy ⊃ $Rt^{\alpha}y$)	∀E(1)	
5. Ht^{α} ⊃ Gt^{α}	∀E(2)	
6. ∀y(Gy ⊃ $Rt^{\eta}y$)	MP(3, 4)	[note how the new constant trumps]
7. Gt^{α} ⊃ $Rt^{\eta}t^{\alpha}$	∀E(6)	
8. Ht^{α} ⊃ $Rt^{\eta}t^{\alpha}$	HS(5, 7)	
9. ∀y(Hy ⊃ $Rt^{\eta}y$)	∀I(8)	[starting from the left]
10. ∃x∀y(Hy ⊃ Rxy)	∃I(9)	

b

1. ∃x∃yRxy	AP [for conditional proof]	
2. ∃yRay	∃E(1)	[starting from the right]
3. Rab	∃E(2)	
4. ∃xRax	∃I(3)	[starting from the left]
5. ∃y∃xRyx	∃I(4)	
6. ∃x∃yRxy ⊃ ∃y∃xRyx	CP(1-5)	

In the following examples, we have the opportunity to apply uses of the methods we called Conditional Proof and Indirect Proof when we presented the Natural Deduction proof system for the standard Propositional Logic, PROPnd*, which we have extended to PREDnd. The sequents we prove are important for suggesting ways in which all nested quantifier symbols can be extracted from within parentheses and placed as iterated quantifier symbols in front – to generate what is known as the Prenex Form of a well-formed formula. (We also keep in mind that empty-premise-set sequents establish tautologies.) Given the liberalization of the rule about registering the superscript symbols for new and arbitrary constants, we may omit those indicators. Other sequents that are also seminal in forming the Prenex Forms are presented subsequently as exercises.

1. ∀x(Fx ⊃ p) P /.. ∃xFx ⊃ p

↱2. ∃xFx	AP
3. Fa	∃E(2)
4. Fa ⊃ p	∀E(2)
↳5. p	MP(3, 4)
6. ∃xFx ⊃ p	CP(2-5)

1. ∃xFx ⊃ p	P /.. ∀x(Fx ⊃ p)
↱2. ~ ∀x(Fx ⊃ p)	AP/IP
3. ∃x ~ (Fx ⊃ p)	∀∃(2)
4. ∃x ~ (~ Fx ∨ p)	Implication(3)
5. ∃x (~ ~ Fx · ~ p)	DeMorgan(4)
6. ∃x(Fx · ~ p)	DN(5)
7. Fa · ~ p	∃E(7)
8. Fa	Simpl(7)
9. ~ p	Simpl(7)
10. ∃xFx	∃I(8)
↳11. p	MP(1, 10)
12. ~ ~ ∀x(Fx ⊃ p)	IP(2, 9, 11)
13. ∀x(Fx ⊃ p)	DN(12)

III.10 EXERCISES

(1) Detect and identify precisely what errors are committed in the following fallacious proofs. Also fill in the justification lines in the right, beneath the indications for premise-status. Additionally, apply the appropriate superscripts in the main lines of the proof. Finally, give intuitive semantic instances of the incorrect proof.

1. ∀x∃y(x ≠ y)	P
2. ∃y(a ≠ y)	
3. a ≠ a	
4. ∃x(x ≠ x)	

1. ∀x(Fx ⊃ Gx)	P1
2. Fa	P2

3. $Fa \supset Ga$	
4. Ga	
5. $\forall xGx$	

1. $\exists xFx$	P1
2. $\exists xGx$	P2
3. Fa	
4. Ga	
5. $Fa \cdot Ga$	
6. $\exists x(Fx \cdot Gx)$	

1. $\forall x(Fx \lor Gx)$	P
2. $Fa \lor Ga$	
3. $Fa \lor \forall xGx$	
4. $\forall xFx \lor \forall xGx$	

1. $\forall xRxx$	P
2. Rab	
3. $\exists yRay$	
4. $\exists x\exists yRxy$	

1. $\exists xRxx \supset \exists ySyy$	P
2. $Raa \supset \exists ySyy$	
3. $Raa \supset Sbb$	
4. $\exists y(Raa \supset Syy)$	
5. $\exists x\exists y(Rxx \supset Syy)$	

1. $Rf(a)f(b)$	P1
2. $a = b$	P2
3. $Rcf(b)$	
4. Rcd	
5. $c = d$	
6. Rcc	
7. $\exists xRxx$	

(2) We have adapted the system PROPnd* from section VIII.15 to construct PREDnd with the added rules we have stipulated for for quantifier symbols, identity and function symbols. We will now present certain exercises about how to use the system.

- Identify the rule(s) of PREDnd that has been applied to justify the line that is generated from the given lines.

a. ∀xFx ⊃ ∀xGx, ~ ∀xFx => ~ ∀xFx
b. ∀x(Rf(x)x ∨ Rm(x)f(x)) => Rf(a)a ∨ Rm(a)f(a)
c. ∃xFx ⊃ ~ ~ ∃yGy => ∃xFx ⊃ ∃yGy
d. (Fa · Gb) ∨ (Fa · Hb) => Fa ∨ (Gb · Hb)
e. ~ ∀x(Fx ⊃ Gx) => ∃x ~ (Fx ⊃ Gx) => ∃x ~ (~ Fx ∨ Gx) => ∃x(~ ~ Fx · ~ Gx) => ∃x(Fx · ~ Gx)
f. Rf(a)m(b) => ∃yRf(a)y
g. ∀x∃yRxy ≡ ~ ∃x∃ySyx => (∀x∃yRxy ≡ ~ ∃x∃ySyx) ∨ Rab => Rab ∨ (∀x∃yRxy ≡ ~ ∃x∃ySyx) => ~ Rab ∨ (∀x∃yRxy ≡ ~ ∃x∃ySyx) => Rab ⊃ (∀x∃yRxy ≡ ~ ∃x∃ySyx)
h. ∀x∀yRxy · ∀x∀ySxy => ∀x∀ySxy · ∀x∀yRxy

- Apply the indicated rule(s) to generate the indicated line from the given lines.

    a. ∀x(Rxa ≡ Rbx), Rba => [∀E] => [MP] =>
    b. ∃y(Fy ⊃ ∀xRxy) => [∃E] => [∃I]
    c. ∀x∀y(Rxy ⊃ ~ Ryx), Rcd => [∀E] => [∀E] => [Contraposition] => [DN] =>
    d. Fa => [∃I] => [∀∃]
    e. ∃x∀y(Rxy ⊃ x = y) => [∃E] => [∀E] => [CE] =>
    f. Rs(a)s(b) => [∃I] => [∃I]
    g. ~ (∀xFx ⊃ ∃xFx) => [CE] => [DeM] => [∀∃] =>

(3) Which of the following are instances of correct – and which are instance of incorrect – rule application?

a. ∃x(Fx · Gx) => Fa · Gb
b. ∃xRxb => Rbb
c. ∀x∀yRxy => Raa
d. ∃x∃yRxy => ∃yRay

e. Fa ⊃ (Ga ≡ ~ Fa) => ∃x(Fx ⊃ (Gx ≡ ~ Fx))
f. ∀x∃yRxy => ∀xRxb
g. ∃x∀yRxy => ∀yRay => Raa
h. ∀x∃yRxy => ∃yRay => Rab
i. ~ (Fa ∨ Ga) => ∃x ~ (Fx ∨ Gx) => ∀x (~ Fx · ~ Gx)
j. ∀x∀y(x = y ⊃ Rxy) => ... => a = a ⊃ ∃x∃yRxy
k. Fa ⊃ Ga => ∃x(Fx ⊃ Gx) => ... => ∃x ~Fx ∨ ∃xGx => ... => ∀xFx ⊃ ∃xGx
l. Ff(b) => ∃xFx
m. (f(a) ≠ f(b)) => ... => ∀x(Rxx ⊃ Rf(a) ⊃ ~ Rf(b))
n. ~ ∃x(Fx · Gx) => ... => ~ (Fa · Ga)

(4) Prove the following sequents in PREDnd.

a. ∀x(Fx ∨ p) /.. ∀xFx ∨ p
b. ∀xFx ∨ p /.. ∀x(Fx ∨ p)
c. ∃x(Fx · Gx) /.. ∃xFx · ∃yGy
d. ∀xFx ∨ ∀xGx /.. ∀x(Fx ∨ Gx)
e. ∀x(Fx ≡ Gx) /.. ∀xFx ≡ ∀xGx
f. ∀x(Fx ∨ Gx) /.. ∃xFx ∨ ∀xGx
g. ∃x(Fx ⊃ p) /.. ∀xFx ⊃ p
h. ∀xFx ⊃ p /.. ∃x(Fx ⊃ p)
i. ∀x(Fx ≡ p) /.. ∃xFx ≡ p
j. p ⊃ ∀xFx, ∃xFx ⊃ p /.. ∀x(Fx ≡ p)
k. ∀xFx ⊃ ∀yGy /.. ∃x(Fx ⊃ Gx)
l. ∃xFx /.. ∀x(Fx ⊃ p) ⊃ p
m. ∃x∃y(Rxy ∨ Ryx) /.. ∃x∃yRxy

IV. SEMANTIC TREES FOR PREDICATE LOGIC: $\mathfrak{T}_{PRED}$

We extend the semantic tree method for Propositional Logic $\mathfrak{T}$ with the addition of tree rules for the two new (finitary-domain) connectives of predicate logic, which are the universal and existential quantifier symbols, and rules for identity and for function symbols. We can use interchange rules for the quantifier symbols as we stipulated them in the natural deduction system of the preceding section. Usually, no tree rules are given for the negated quantifier symbols - with reliance, instead, on the afore-mentioned interchange rules. We could, however, supply such rules - which seems an elegant move to make considering how the propositional connectives rules for the semantic tree come in these two varieties: rules for the connectives and for the negated connective symbols (with the exception of the negation symbol which does not have a tree rule and is considered unresolvable by tree rules - signifying that the formula within its scope is valuated as false.) The implementation of the semantic tree procedure for predicate, extending the system $\mathfrak{T}$, can be named $\mathfrak{T}_{pred}$.

In the parallel trees method, as we know from 0.4.3, we construct separate trees for the premises set and for the conclusion. On termination, and possibly after saturation of the premises tree, we determine validity or invalidity of the given argument form as follows:

The argument form is valid if and only if every premise set is covered (in a technical sense to be made precise) by at least one conclusion set.

The symbol "⊇" is used standardly to denote supersethood, with the (not necessarily proper) superset written to the left of the symbol.

The premise and conclusion sets are populated by the truth values of the atomic letters up the open paths. A set X is covered by a set Y if and only if X is a superset of Y: as such every member of Y is in X. The set X, the superset, may or may not have additional members relative to those in Y – in other words, it does not have to be a proper superset. Not every conclusion set needs to be involved in the tally; but every premise set (possibly after saturation) must be covered for the determination to be that the argument form is valid. Vacuous and trivial entailment are straightforwardly accommodated. Trivial entailment (if we may use this word in the case of the standard logic) occurs in the case in which the conclusion is a tautology – and, as such, validly entailed by any set of premises. Vacuous entailment encompasses the case in which a contradiction validly entails any conclusion whatsoever. Of special interest is the case of the closed premises tree; the closed tree is not the same as an empty set but should rather be thought of as not a set at all. A closed tree is to be considered as covered and we can say that this covering is precisely a vacuous case. (Of course, the empty set is not a superset of any set other than itself.)

We make a stipulation: If the given well-formed formula is not in prenex form, as given, we should convert to an equivalent formula that is in prenex form. We could make other stipulations instead of this one but it is easier to take this approach.

For the predicate inferential schemata, we need metalinguistic symbols for the new resources of predicate logic: individual variables and individual constants (the terms), predicate constants and, of course, the quantifiers; also for identity and for function symbols. We stipulate use of the same symbols, for the sake of convenience, except that we use geometrical shapes for symbolizing propositional formulas (not necessarily atomic) in the rule-schemata. There is something to pay attention to in this case. The propositional formula within the scope of a quantifier symbol is presented without showing the individual variables in it. Thus, it is presented unanalyzed – as in Propositional Logic. It is strictly presumed that the individual variable accompanying the quantifier symbol is *free in the formula* before the quantifier symbol was applied.

The rule for the universal quantifier, for instance, compels removal of the quantifier symbol and uniform substitution of any – not necessarily a new – individual constant for that quantifier's variable. As in the case of constructing natural deduction systems, "new" means for our purposes not yet appearing in any proof line up to the point of applying the rule we are discussing. For the existential quantifier rule, the move compelled by the tree rule requires removal of the quantifier symbol and uniform substitution for the quantifier's variable of a new individual constant. We need to indicate schematically the differentiation between any constant and a constant that has to be "new" in the sense specified above. Texts sometimes "flag" this new constant – in the body of the actual proofs – and we need some specific symbol for the inferential schemata agreeing – for the sake of convenience – to use the same symbol in the implementation of the rules within actual proofs. We use the superscript "^{n}" to indicate a new constant.

We recapitulate the propositional connectives tree rules for convenience and we add the predicate rules – for the quantifier symbols and the negated quantifier symbols – before we introduce whatever additional restrictions may be needed. It is understood that the structural rules for the semantic tree method for Propositional Logic apply.

Vertical Rules: ~~R, ·R, ~ ∨R, ~⊃R

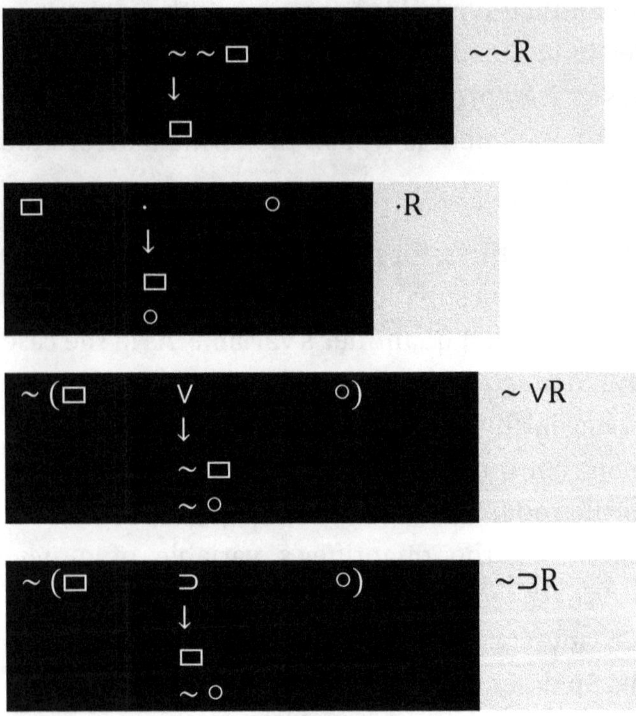

Branching Rules: ∨R, ⊃R, ~·R, ≡R, ~≡R

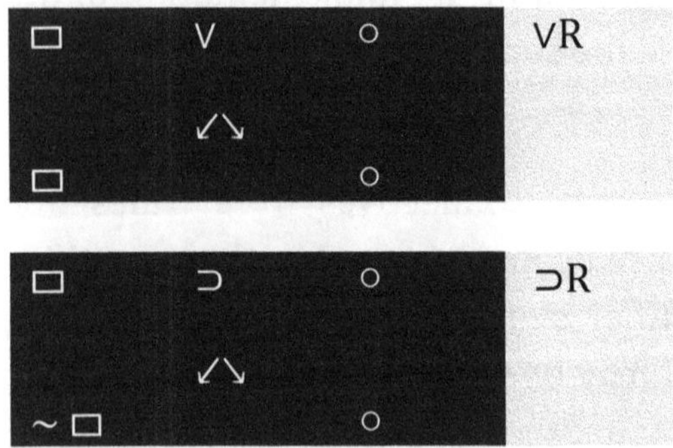

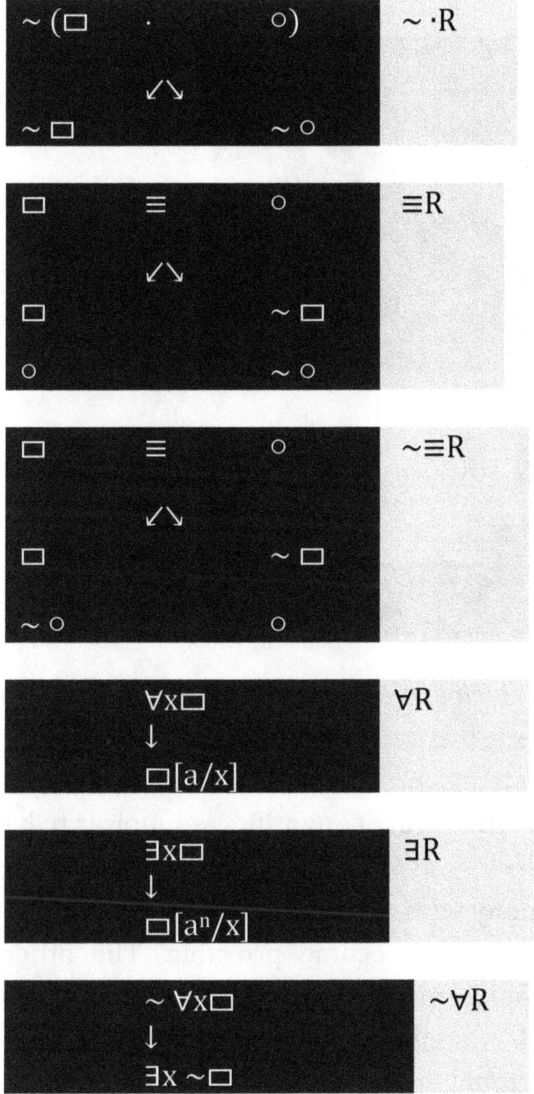
~ (□ · ○)
~ □ ~ ○
~·R
□ ≡ ○
□ ~ □
○ ~ ○
≡R
□ ≡ ○
□ ~ □
~ ○ ○
~≡R
∀x□
↓
□[a/x]
∀R
∃x□
↓
□[a^n/x]
∃R
~ ∀x□
↓
∃x ~□
~∀R

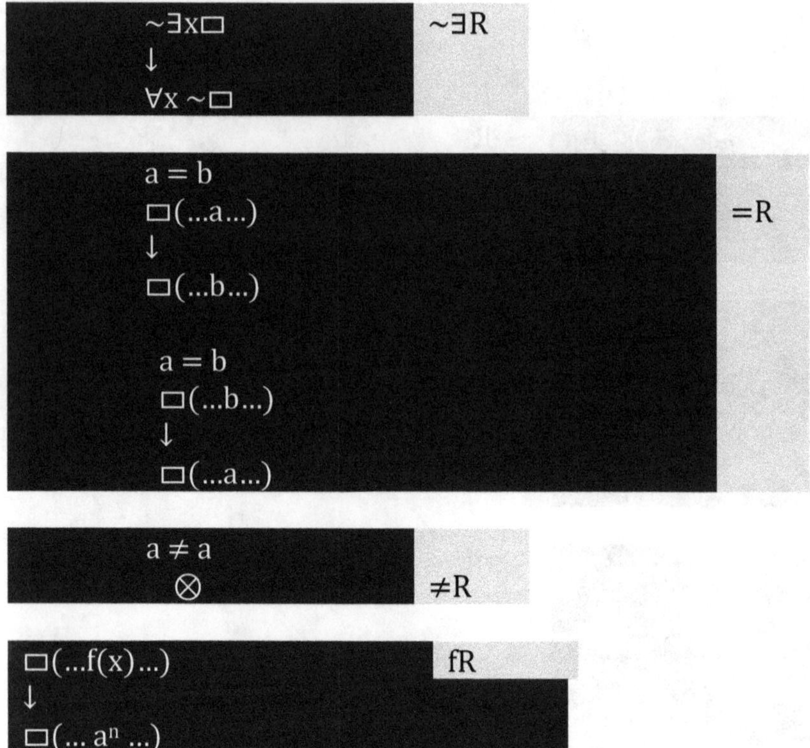

The rules for the sequence in which nested quantifiers are managed in eliminations and introductions are as in IX.11. This is important since predicate logic decision systems present additional challenges pertaining to the order in which removal of quantifier symbols is to be conducted. Violation of the rules, as we saw in IX.11, results in dismantling the system's harmony with the semantics of the predicate logic because invalid sequents then become provable. The other stipulations about uniform replacement of the variables when the controlling quantifier symbol is removed also apply in this case. Restrictions that have been stipulated, if they apply, are considered imported into $\mathcal{T}_{pred}$.

The procedures for checks are as in the tree method for Propositional Logic. The check for validity/invalidity proceeds by attaching the negated conclusion to the root, along with the given premises; validity is established if and only if all paths of the tree are closed - which means that the tree itself is declared closed. If any path is open, that means that the given argument form is invalid and the values for the atomic components (the counterexample or countermodel values) are read off the open path. The valuations we extract show us extensions of predicate letters and referents of individual constants, based on truth and falsehood of formulas: for instance, ⌜ Fa ⌝ being an atomic propositional formula, its truth on an open tree path indicates that the object denoted by ⌜ a ⌝ belongs to the extension of the predicate letter denoted by ⌜ F ⌝. Thus, we can extract model information on the basis of the completed semantic tree.

The checks for the logical status of a formula and for consistency or co-satisfiability of a given set of formulas proceed as in the case of semantic trees for Propositional Logic.

The inequality rule compels closing the branch - and associated path - as soon as we have a line with inequality between any letter and itself. It is taken as indisputably a matter of logical absurdity to claim that something is not identical with itself. The law of self-identity is considered fundamental. More precisely speaking, an inequality line should be read, within our formalism, as: "a does not co-refer with itself" or "a has a unique referent that is distinct from its ... unique referent." Regarding function symbols, the function-rule gives us permission to replace a function symbol by means of an individual constant that is new. We may or may not decide to make use of this rule depending on the strategy required for a proof. We could also add a rule that, as shown below, permits replacing a function by the regimentation uniqueness-formula with the regimentation predicate letter ⌜ F ⌝. [Is this rule provable from the other rules we already have in the system?]

□(...f(x)...) (f/∃∀)R
↓
□(...∃x∀y(Fy≡x=y)...)

IV.2 A POSITIVE-TREE SEMANTIC TREE METHOD FOR PREDICATE LOGIC: $\mathrm{T}_{\mathrm{PRED}||}$

We extend the Propositional Logic paired-trees decision procedure of 0.4.3, $\mathrm{T}_{||}$, to a predicate paired-trees system, $\mathrm{T}_{\mathrm{pred}||}$. Accordingly, our metalinguistic symbols – which can be superimposed on the tree diagram – are "⊗" for a closed branch (and, hence, the path to which the branch belongs); open paths, upon termination, are marked underneath by "⊕". An additional symbol we utilized in $\mathrm{T}_{\mathrm{pred}||}$ was "◉" used to indicate provisional termination. This reminds us of the amenity we have in the system of continuing even after termination, by adding splitting branches for any atomic letter ⌜ p ⌝ and its negation ⌜ ~ p ⌝. This is called saturation and is needed often before we can pronounce the final verdict. The explanation is straightforward: appending on an open path corresponds to conjunction; the splitting branches that are appended trace an inclusive disjunction with the disjuncts labeling the added splitting branches. The values we read on the tree paths are not affected; the rationale can be read off the following standard logic validity:

$$(\varphi \cdot (\psi \vee \sim \psi)) \equiv \varphi$$

The provisions for $\mathrm{T}_{\mathrm{pred}}$ apply also in the case of $\mathrm{T}_{\mathrm{pred}||}$. The structural rules and the connectives rules are the same. Indeed, checking for consistency of a set of formulas by the parallel trees method is not different from that of the negative tree system: although there is no conclusion to compel construction of a parallel tree, the method may well be used for a consistency check.

Matters turn rather interesting when we repair to the accommodation of quantifier rules in $\mathcal{T}_{pred||}$. Notice how the valid inference below, which is merely a relettering, cannot be proven if we stick to the $\mathcal{T}_{pred||}$ rules for the existential quantifier. Thus, ∃R will be found in need of amendment.

∃xFx ⊢ ∃yFy

1. ∃xFx ↓ Fa^n — 1. ∃yFy ↓ Fb^n

Premise Set: P = {Fa}
Conclusion Set: C = {Fb}
P ⊉ C

Saturation will not resolve the issue.

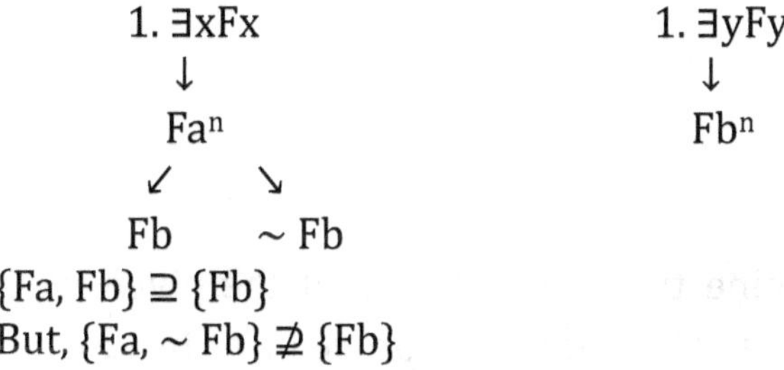

{Fa, Fb} ⊇ {Fb}
But, {Fa, ~ Fb} ⊉ {Fb}

The rules for the unnegated quantifiers, ∀R and ∃R, are amended relative to $\mathcal{T}_{pred}$. All other rules remains as they are in $\mathcal{T}_{pred}$. The emendation compels us to expand those quantifiers so that – as is semantically justified – the existential quantifier is treated as an inclusive disjunction – and, as such, compels generation of splitting branches – and the universal quantifier is a conjunction – and compels generation of vertical branches for the premises and for the conclusion trees. The added branches are for all the individual constants of a finitary domain. The general case can be shown as below, also showing how the argument form contemplated above is checked appropriately as valid. Our grammar allows us to use subscripts for individual constant letters. Also, when it comes to inclusive disjunction - and given that it is associative - we may generate splitting branches at once underneath an open branch.

1. ∃xFx   1. ∃yFy
↙… ↓ …↘   ↙… ↓… ↘
$Fa_1 \cdots Fa_j \cdots Fa_n$ $Fa_1 \cdots Fa_j \cdots Fa_n$
◉…◉…◉   ◉…◉…◉

Every path of the premises set is covered by some conclusion set.
We show another example.
∃xFx ∨ p $\top_{pred||}$ ⊢ ∃x(Fx ∨ p)

1. ∃xFx ∨ p   1. ∃x(Fx ∨ p)
↙↘   ↙… ↓… ↓ ↘
∃xFx   p $Fa_1 \cdots Fa_j \cdots Fa_n \cdots p$
↙…↓…↘   ◉
$Fa_1 \cdots Fa_j \cdots Fa_n$
◉…◉…◉

IV.3 EXAMPLES

1. We are asked to determine the logical status of the given well-formed formula by means of the $\top_{pred}$ decision procedure. In addition to implementing the procedure, and before we do so, we will show how we can reflect insightfully on whether the given formula expresses a tautology, or contradiction, or not. We have two options to apply to assist our efforts. We may convert the given formula to an equivalent prenex form; or we may reverse the steps that generate a prenex form to yield an equivalent form that is not prenex. We may take such steps

because the resultant form may afford us insights into the status of the formula. If, for instance, there can be a falsification instance, we know that we are not dealing with a logical truth or tautology. We may also resort to the semantic or model-constructing approach of I.5-9.

Another initial intervention we may make is to attempt to render an intuitive translation of a proposition that has the form expressed by the given formula (before or after conversions relative to the prenex form.) These two initial steps are informal and may or may not shed insight. They are not to be trusted as warranting decision on intuitive grounds but they may assist our understanding and hone reasoned expectations.

$\mathfrak{T}_{pred} \vdash \forall x(Fx \supset \forall xFx)$

a. Extracting an equivalent prenex form yields the following:

$$\forall x \forall y(Fx \supset Fy)$$

Translating (a proposition that has this form) in a step-by-step pedantic fashion (not in idiomatic English), we have: "For all x and all y, if x Fs then y Fs too." We know that differentiation of variables is required for the conversion to prenex: thus, we don't use the same variable, for instance ⌜x⌝, in the prenex. Hence we have the proposition of the sentence above. This appears falsifiable. Using, metalinguistically, modes of speech that allow us to quantify over predicate letters themselves: we have the statement that all F-predicates are such that for everything, if it Fs, then everything else Fs as well.

The formula is equivalent to the following non-prenex formula.

$$\exists xFx \supset \forall yFy$$

Reading this, we spell out the rather implausible inference from some entity's having the F-property to the claim that everything has the F-property. This can be falsified by means of the following countermodel, which we can construct after we have obtained the expansion of the formula for a two-member domain – the least number for which a countermodel can be obtained.

Exp(∃xFx ⊃ ∀xFx) = ((Fa ∨ Fb) ⊃ (Fa · Fb))

$\mathcal{M}$ = <D, √>
D = {①, ②}
√(①) = a
√(②) = b
√(F) = {①}
√(a) ∈ √(F) ⇒ √(Fa) = T
√(b) ∉ √(F) ⇒ √(Fb) = F
√(Fa ∨ Fb) = √(T ∨ F) = T
√(Fa · Fb) = √(T · F) = F

b. Now we apply the decision procedure of our tree system to the given formula. We begin with observing the stipulation requiring conversion to an equivalent prenex form.

1. ~ ∀x∀y(Fx ⊃ Fy)	
↓	
2. ∃x ~ ∀y(Fx ⊃ Fy)	~∀R(1)
3. ∃x∃y ~ (Fx ⊃ Fy)	~∀R(2)
4. ∃y ~ (Fa ⊃ Fy)	∃R(3)
5. ~ (Fa ⊃ Fb)	∃R(4)
↓	
6. Fa	~⊃(5)
7. ~ Fb	~⊃(5)
⊕	
⊬ ∀x(Fx ⊃ ∀xFx)	

The single tree path remains open, which shows that the given formula is not a tautology (since its negation is not a contradiction.) We can extract countermodel values from the open path.

<u>Falsifying Model</u>

$\surd(Fa) = T \Rightarrow \surd(a) \in \surd(F)$

$\surd(\sim Fb) = T \Rightarrow \surd(Fb) = F \Rightarrow \surd(b) \notin \surd(F)$

M = <D = {❶, ❷}, √ = {{<a, ❶>, <b, ❷>}, F = {❶, ❷}}>

2. $\exists xRxx \vdash_{Tpred} \exists x\exists yRxy$

a. Analyzing, we have the following: assume that there is anything – at least one thing – that is R-related to itself; then, it must be that there is at least an x and there is at least a y such that x is R-related to y. Note that it is not specified – by means of a restriction – that x and y cannot be identical. It is rather like saying that, if something is related to itself, it follows that something is related to something without excluding the identity of the presumed things that are related to each other.

b. If we develop expansions, for semantic modeling, we have for a two-member domain.

$Raa \vee Rbb \vdash Raa \vee Rab \vee Rbb \vee Rba$

There seems to be no way we can make the premise true and the conclusion false. This must be a valid argument form.

The verdict is different if it is stipulated that the, at least two, things related, mentioned in the conclusion, cannot be identical. Then, the rendering is: Assume that at least one thing is R-related to itself; therefore, there are at least two things such that one is R-related to the other. The development by expansion discloses possibilities or options for making the premise true and the conclusion false. We leave this as an exercise.

b. We apply the decision procedure to the given logical form and to the logical form we discussed above.

1. ∃xRxx /.. ∃x∃yRxy	
2. ~ ∃x∃yRxy	negated conclusion
3. ∀x ~ ∃yRxy	~ ∃R(2)
4. ∀x∀y ~ Rxy	~ ∃R(3)
5. Raa	∃R [a^n]
6. ∀y ~ Ray	∀R [a]
7. ~ Raa	∀R [a]
⊗	

∃xRxx ⊢$_{Tpred}$ ∃x∃yRxy

The tree is closed, which shows that the given argument form is valid.

1. ∃xRxx /.. ∃x∃y(x ≠ y · Rxy)		
2. ~ ∃x∃y(x ≠ y · Rxy)		negated conclusion
3. ∀x ~ ∃y(x ≠ y · Rxy)		~∃(2)
4. ∀x∀y ~ (x ≠ y · Rxy)		~∃(3)
5. Raa		∃R(1)
6. ∀y ~ (a ≠ y · Ray)		∀R(4)
7. ~ (a ≠ a · Raa)		∀R(5)
↙	↘	
8. ~ a ≠ a	~ Raa	left: ~· R(7) right: ~· (7)
	⊗	
9. ~ ~ a = a		left: def(≠)(8)
10. a = a		left: ~~R(9)

∃xRxx ⊬ ∃x∃y(x ≠ y · Rxy)

There is a Falsifying Model (or Countermodel.)

One path remains open. Interestingly, our countermodel has only one member that is related to itself. The conclusion form is instantiated by the claim that "there are at least two things one of which is related to the other." This is falsified when there are not at least two things. Thus, it is falsified if there is at most or exactly one thing. This means that the conclusion is false. If the existing thing is R-related to itself, then the premise proposition is true. Hence, we have a countermodel.

<u>Falsifying Model</u>

$Raa \nvdash a \neq a \cdot Raa$

$M = \langle D = \{❶\}, \surd = \{\langle a, ❶\rangle, R = \{❶, ❶\}\}\rangle$

IV.4 EXERCISES

(1) Check by means of the semantic tree method procedure Ʈ$_{\text{pred}}$ if the following sequents are valid or invalid. (Alternatively, we may think of this enterprise as consisting in checking the logical status of the formulas to the right of the turnstile if there are no formulas to the left of the turnstile: this is the case of empty-premises logical consequences.) If invalid, produce the countermodel values of the atomic variables.

a. $?\vdash \forall x\exists y(x = y)$
b. $?\vdash \exists x(\exists yFy \supset Fx)$
c. $?\vdash \forall x\forall yRxy \supset \forall y\forall xRxy$
d. $?\vdash \forall x\forall yRxy \supset \forall xRxx$
e. $?\vdash \forall x\exists yRxy \supset \exists y\forall xRxy$
f. $?\vdash \exists x\forall yRxy \supset \forall y\exists xRxy$
g. $?\vdash \exists x\exists yRxy \equiv \exists y\exists xRxy$
h. $?\vdash \exists x\exists y(x \neq y \cdot Rxy) \supset \exists xRxa$
i. $?\vdash \forall x(Fx \supset \exists xFx)$ [convert to prenex form]
j. $?\vdash \exists x(Fx \supset \forall xFx)$ [convert to prenex form]
k. $\exists x\exists yRxy\ ?\vdash \exists x\forall yRxy$
l. $?\vdash \forall x(Fx \supset Gx) \supset (\forall xFx \supset \forall xGx)$
m. $?\vdash \exists x(Fx \vee Gx) \equiv (\exists xFx \vee \exists xGx)$
n. $\forall xFx \vee \forall xGx\ ?\vdash \forall x(Fx \supset Gx)$
o. $?\vdash \forall x(Fx \cdot Gx) \equiv (\forall xFx \cdot \forall xGx)$
p. $\forall x(Fx \equiv Gx)\ ?\vdash \forall xFx \equiv \forall xGx$
q. $\forall x(p \equiv Fx)\ ?\vdash p \equiv \exists xFx$
r. $\forall xFx \supset \forall yGy\ ?\vdash \exists x(Fx \supset Gx)$

s. $?\vdash (\forall xFx \supset p) \supset \exists x(Fx \supset p)$
t. $\exists x(Fx \supset p)\ ?\vdash \forall xFx \supset p$
u. $\exists xFx\ ?\vdash \forall x(Fx \supset p) \supset p$
v. $\exists x\exists y(Rxy \vee Ryx)\ ?\vdash \exists x\exists yRxy$
w. $?\vdash a = b \supset (Rf(a)f(b) \supset \exists xRxx)$

(2) Study again the translations of properties of relations and then use the semantic truth method to check if the following are true or not.

a. (Reflexive + Euclidean) => Convergent
b. Symmetrical => Convergent
c. One-Step Dead-End => Dense
d. (Symmetrical + Transitive) => Euclidean
e. Reflexive => $\exists x(Rax \cdot Rxa)$
f. Euclidean => Transitive
g. (Reflexive + Euclidean) => Symmetrical
h. Convergent => Reflexive
i. (Anti-symmetric + Reflexive) => Connected

(3) Check the following for validity and discuss. First, expand the uniqueness quantifier:

$$\boxed{\exists!xFx \equiv \exists x\forall y(Fy \equiv x = y)}$$

a. $?\vdash \forall x\exists!y(f(x) = y)$
b. $?\vdash \forall x\exists!y(f(y) = x)$
c. $?\vdash \exists x\exists y(f(x) = y)$
d. $?\vdash \forall x\exists y(f(x) = f(y))$

(4) Do the preceding exercises by applying the paired trees method for predicate logic, $\mathcal{T}_{pred||}$.

V. APPENDIX ON SET THEORY, MATHEMATICAL FUNCTIONS AND RELATIONS

A set is a collection of objects that are all specified in some appropriate fashion so that there is no ambiguity about what is included and so that, of any specified object, we can say if that object is included or not. The objects that are in the set are called members or elements of a set. The members of a set may or may not be related to each other by any shared property. (There is a view, to be found in some, especially older, texts, that considers general collections of objects as classes and reserves the term "set" for collections whose members share some property.) What is important for having a set is that it has been specified completely and unambiguously what the elements or members of the set are. The use of enumeration is not the only way to define a set, as we will see, but any set is presumed to be so specified that it is precise, for any suggested item, whether the item is included in the set or not. Sets need not have a finite number of members. For instance, the set of natural numbers has an infinite number of members. Given our definition of a set, it is necessary that we can tell about any item, whatsoever, whether it belongs to a set or not.

It may be thought a stretch, or illegitimate, to accept infinitarian sets since we cannot verify inclusion or exclusion of all members procedurally. There is a school of Mathematics, Intuitionism, which accepts only mathematical objects that can be constructed, at least in principle; this is in contrast with the standard approach, which accepts any abstract object, regardless of verifiability or constructability of the object, insofar as assuming it does not lead to a contradiction. Intuitionists accept infinite sets that are called countable or denumerable: this means that the members of such sets can be put in one-to-one correspondence with the members of the set of the natural numbers. This size of infinity (with countability being, obviously, a special set-theoretic concept) is considered to be actual infinity by Intuitionists who accept it but they reject potential infinity – which is any size such that it cannot be put in one-to-one correspondence with the members of the set of the natural numbers.

We use the symbols "{" and "}" as left and right fishhooks to enclose the symbols representing the members of a set and we use capital letters, possibly with subscripts, for labeling sets. What symbols are used for set members depends on what kinds of objects the members are - if numbers, numerical symbols are used and small letters from the alphabet may be used for other objects. The names of the sets themselves can be, instead of capital letters, whole words of English – although this is a rather rare practice. This is all a matter of convention but, in general, there is no reason to impose prior restrictions on what symbols may be used for labeling set members or the sets themselves.

There are three ways in which a set can be defined or characterized:

a. Extensionally or by full enumeration or representation of its members. Obviously, it is not the members themselves that we write but conventional labels or symbols for the members.

We can have heterogeneous elements as members of sets. For instance, in this set the members – as is obvious from an intuitive reaction to the symbols – are of different kinds. One member is a number (with its name being the numeral "2"), another is a letter of a Latin-based alphabet and the third is an arrow.

X = {2, a, ↘}

The order in which the member symbols are written is not important. If the order is changed, we still have the same set insofar as all the objects – and the same objects – are represented as being included.

Repeating the symbol of a set member does not change anything because two sets with exactly the same members are defined as being identical with each other. Identity between any two sets is symbolized by "=".

Thus, the following two extensional representations of members represent the same set.

{1, 2, 3, 1} = {1, 2, 3}

Since the order in which the elements are presented is also immaterial, we also have, as an example:

{2, 3, 1} = {1, 2, 3} = {1, 3, 2} = {2, 1, 3}

Since the order in which the member symbols are arranged is irrelevant to the characterization of a set, for the purpose of depicting set-theoretic arrangements in which the order *is* important we will define soon another concept that is called <u>ordered pair</u> (for two ordered items) or, generally, <u>ordered n-tuple</u> (for n items that have specific order imposed on them.)

For infinite sets, notice the device deployed below for the set I^+ of positive integers.

Examples of sets defined extensionally:

A = {a, b, c}
B = {1, 2, a, b}
I^+ = {1, 2, ...}
X = {♚, ♛, ♜, ♝, ♞, ♟}
Z = {\, -, /}
EVEN = {2, 4, 6, ...}
M = {{{}}, {}, {1}, {0}}
X = {φ, ψ, ...}
C = {~, ·, ∨, ⊃, ≡, ≢}

b. Sets can also be defined <u>intensionally</u> or <u>by abstraction</u>: this is done by specifying the property shared by all the set members, if there is such a property.

Examples of intensional definitions of sets are given below. We also give the extensional definition for the same set. Notice how a set with infinite members, like the set of odd numbers O, is represented extensionally.

A = {x/ x is an integer equal to or greater than 5} = {5, 6 ...}

B = {x/ x is a student in class X in college Y in semester Z}
WFF = {x / x is a well-formed formula of a Propositional Logic idiom}
X = {x/ x is a symbol of a chess piece} = {♚, ♛, ♜, ♝, ♞, ♟}
O = {x/ x is an odd number} = {1, 3, 5, 7, ...}

If more than one property is involved, we have a way of writing the set by abstraction, as we will now see. It should be noted that any object whatsoever has the property of being self-identical. So, at the very least, this property is possessed by any object. The set is to be considered all its members: this much is obvious. Now, when it comes to contemplating whether some item symbolized by "a" belongs to the set X in terms of some property the item has, this is what we say:

An item a belongs to the set X if and only if a has property F1 or ... or a has property
[Fn, where these all the properties that ensure membership in X.]

This "either-or" in the definition has the meaning of inclusive disjunction, with which we are familiar from our study of Propositional Logic. "Either φ or ψ" is true, in the inclusive sense, if and only if at least one of the propositions denoted by "φ" and "ψ" is true. This includes the case in which both are true.

The following examples illustrate the case of definition of a set by abstraction.

A = {x/ x = 1 or x = 2 or x = a or x = b}
B = {x/ x is a piece of chess}
P = {x/ x is a prime number}
X = {x/ x is a President of the United States up to and including Ronald Reagan}
Z = {x/ x is the object "\", or x is the object "-", or x is the object "/"}
EVEN = {x/ x is an even number}
M = {x/ is any combination of left and right parentheses}
X = {φ, ψ, ...}
C = {x/ x is a connective of the standard Propositional Logic

symbols, such that x is negation or x is conjunction or x is inclusive disjunction or x is material implication or x is equivalence or x is exclusive disjunction} = {~, ·, ∨, ⊃, ≡, ≢}

c. Finally, sets can be defined by recursion or presentation of specified clauses which are the rules that are to be applied in constructing the set and, therefore, in determining of anything whether it is a member of the set or not.
   For example:
   A = {3, 4, 5, ...} = {x/ x is an integer equal to or greater than 3} can be defined recursively as follows:
   Clause 1: 3 is in A (basis clause)
   Clause 2: If x is an integer and x = 3 or x > 3, then x is in A
   Clause 3: Nothing else is in A (closure clause)

We symbolize set membership - for x being a member of X - by "x ∈ X" and non-membership - x not being a member of X - by "∉". Thus,
If X = {1, 3, 5}, then 3 ∈ X and 2 ∉ X.

Sets can have other sets as members. So the expression "X ∈ Y" is not nonsense (as it is saying that a set symbolized by "X" is a member of a set symbolized by "Y"); but it is saying something different from "x ∈ Y" which means that a non-set object symbolized by "x" is a member of a set symbolized by "Y."
It is very important to forestall confusion by comprehending and keeping track of the distinction between being a member of a set and being a subset of a set. A subset Y of a set X is defined as a set all of whose members are members of X (with X possibly having more members that are not in Y, in which case Y is proper subset of X.) A set that is a member of another set, on the other hand, is included in the set as a member – it is not a subset, it is not a separate set.

Special sets are the empty or null set, symbolized by "∅" or "{}" (with another older symbolizations being "Λ" or "0"), and the universal set symbolized by "Ω" (and in older texts by "V" or "1").

The empty set is the set with no members. Since two sets are defined to be identical with one another iff (if and only if) they have exactly the same members, it follows that there can only be one empty set since any two presumed empty sets would have the same members exactly (no members!) and then would have to be identical - or be the same set.
The universal set is the set of all the members specified in a given context. There is only one universal set for each specific context - which we can prove in the same way we proved that there can only be one empty set. The empty set and the universal set are definable by abstraction as follows. We take the opportunity to show an alternative and rather common notation.

$\emptyset = \{x/\ x \neq x\}$ or $\emptyset = \{x.\ x \neq x\}$
$\Omega = \{x/\ x = x\}$ or $\Omega = \{x.\ x = x\}$

A <u>subset</u> Y of a set X is a set all of whose members are in X. If there is at least one member of X that is not in Y, we say that Y is a <u>proper subset</u> of X. We symbolize as: $Y \subset X$. Otherwise, Y is an <u>improper subset</u> of X, which we symbolize by: $Y \subseteq X$.
In the above instances, X is respectively the proper or improper <u>superset</u> of Y, symbolized by "$\supset$" and "$\supseteq$" respectively.

As examples, consider:

$A = \{1, 2\}, B = \{1, 2, 4\}, C = \{1, 2, 3, 4\}, D = \{1, 3, 5, 7\}$

We have:

$A \subset B, A \subset C, B \subset C$

The symbol "$\subseteq$" is also used generically, to indicate that either proper or improper subsethood obtains. So, it is not an error to write, for instance, $A \subseteq B$ in the above case although this conveys less information than what we did in fact write to indicate specifically that the set A is a proper subset of B.
C is not a subset of D; not is D a subset of C. It is not the case that, given any two sets, one must be the subset of the other.
It should be obvious, and it is easily provable that two identical sets are each other's improper subsets:

If $X \subseteq Y$ and $Y \subseteq X$, then $X = Y$

Of course, no set can be its own proper subset. The empty set is each set's subset and is a proper subset of every set that is not itself the empty set.

The following relations all hold true:

$X \subseteq X, X \supseteq X$

$\emptyset \subset X$ (if $X \neq \emptyset$)

$\emptyset \subseteq X$ (no matter what members X has)

$X \subset \Omega$ (if $X \neq \Omega$)

$X \subseteq \Omega$ (no matter what members X has)

The powerset of a set X, symbolized by "$\wp(X)$", is the set of all the subsets of X.

$\wp(X) = \{Y/ Y \text{ is a set and } Y \subseteq X\}$

Notice the important distinction between set membership and subsethood. Given our definition above, Y is a subset of X but it is a member of $\wp$ (X).

$Y \in \wp(X), Y \subseteq \wp(X)$

We have seen that the order in which the members of a set are represented in immaterial to the characterization of a set. The question then arises as to what device can be used to indicate that an order is imposed on set members. For instance, we may wonder how we may symbolize ordered membership in a set that is intensionally defined as:

$M = \{x, y/ x \text{ is the mother of } y\}$ (!)

The above symbolization must be rejected because we have no directions prohibiting writing it as,

$M = \{y, x/ x \text{ is the mother of } y\}$.

But, surely, these are characterizations of two different sets. Or, if we take the extensional definition of M, we have:

$M = \{a, b, ...\} = \{b, a, ...\}$

This is wrong again because the order that constrains the two members (the one being the mother of the other and not the other way round) cannot be indicated.

Here is the answer to this interesting challenge.
Relations that are binary (as is the motherhood relation) can be represented set-theoretically as ordered pairs.
For instance, if y is the mother of x - so, y is motherhood-related to x - this is captured by the ordered pair: $<y, x>$. Now we can present the intensional and extensional definitions of a set like M of the above example.

$$M = \{<y, x> / \text{y is the mother of x}\} = \{<b, a>, ...\}$$

An example of a ternary relation is "x is between y and z." This can be represented by means of an ordered triple: $<x, y, z>$. Generalizing, we can ascend to an n-tuple: $<x_1, x_2, ..., x_n>$.
The following equation applies.

$$<x, y> = \{x, \{x, y\}\}$$

To see this, let us consider the two ordered pairs $<x, y>$ and $<y, x>$ which should in principle be considered as not being identical with one another. Had we used set notation, $\{x, y\}$ and $\{y, x\}$ would have to be identical since they have exactly the same members and the order in which names of members appear in the set does not matter. For ordered pairs, however, we have to make the order count. The set-theoretic notation $\{x, \{x, y\}\}$ accomplishes exactly that.

$$<x, y> = \{x, \{x, y\}\} \neq \{y, \{y, x\}\} = <y, x>$$

Next, we define the so-called Cartesian product of two sets.
The Cartesian product of two sets is constructed by taking all members of the first set as first members of ordered pairs with members of the second set as second member of the pair, and we continue doing this to cover all possible combinations.
We use an example to present this subject informally. There are cases in the study of logic when we need this machinery even at the introductory level.

$$X \times Y = \{1, 2\} \times \{3, 4\}$$

We can use a matrix, drawn as you see below, to give a notion of how we form the Cartesian product of two given sets. Make sure that you remember what we defined above as ordered pairs: the use of "<" and ">" that you see is meant to demarcate ordered pairs.

X/Y	**3**	**4**
1	<1, 3>	<1, 4>
2	<2, 3>	<2, 4>

So, the Cartesian product is a set of ordered pairs:

X x Y = {1, 2} x {3, 4} = {<1, 3>, <1, 4>, <2, 3>, <2, 4>}

It is possible to construct the Cartesian product of a set by itself. For instance,

X x X = {1, 2} x {1, 2} = {<1, 1>, <1, 2>, <2, 1>, <2, 2>}

Obviously, X x Y ≠ Y x X

To construct Cartesian Products of more than two sets we proceed in pairs. For instance,

X x Y x Z = (X x Y) x Z

The cardinality or cardinal number of a set (ℂ(X)) is the number of its members. For instance, if X = {a, 1, b, 2, c, 3, d, 4}, then ℂ(X) = 8. Interesting issues arise regarding the cardinalities of infinite sets. Such issues are of no relevance to our purposes in this text. We can mention that the cardinality of the natural numbers is important and is denumerable, as we say. It was proven that different infinite sets may have different sizes – with the denumerable size of infinity defined by means of a theoretical operation that matches each member of an infinite set with one corresponding natural number. A set like that of the real numbers, for instance, has been proven to have a larger size than that of the natural numbers.

Interestingly, we can give an elegant definition of finite set by using the concept of cardinality we have just introduced: A set is finite if and only if its cardinal number is larger than the cardinal numbers of all its proper subsets.

⊛ Set-Theoretic Operations

We define the following operations on sets. These operations are Boolean - with set theory historically one of the very first fields of application of the Boolean algebra. We indicate the Boolean character of the set-theoretic operations by means of certain properties of those operations which we cite below. After you have familiarized yourself with the definitions of the connectives or truth-functions of Propositional Logic, you should revisit this section and try to determine how set-theoretic operations map on corresponding truth-functions of Propositional Logic.

The complement of a set X is the set of all members of the defined universal set that are not in X. This is a unary operation on sets.

$X' = \{x \in \Omega / x \notin X\}$

Compare this to the propositional operation of negation.

A complement of a set X relative to a set Y is the set of all members of Y that are not in X. This is also called the difference of X from Y. (So, the unary complement of X can be defined also as the complement of X relative to the universal set.)

$X_Y = Y - X = \{x \in \Omega / x \in Y \text{ and } x \notin X\}$

$X' = \Omega - X$

The union of two sets X and Y is defined as the set whose members include all the members of X and all the members of Y (with each member written only once, of course.) We can also say that the union of X and Y contains a member x of the universal set iff x either is a member of X or it is a member of Y or it is a member of both X and Y. This recalls the definition of the propositional operation we call inclusive disjunctive. If we use "V" as the symbol for inclusive disjunction, we may write:

$X \cup Y = \{x \in \Omega / \text{either } x \in X \text{ or } x \in Y\} = \{x \in \Omega / (x \in X) \vee (x \in Y)\}$

The intersection or overlap of X and Y is the set whose members include only those members of the universal set which are both in X and in Y. We can also say that a member x of the universal set is in the intersection of X and Y iff it is a member of both X and Y. This evokes the propositional operation of conjunction – which we can symbolize by "".

$$X \cap Y = \{x \in \Omega / x \in X \text{ and } x \in Y\} = \{x \in \Omega / (x \in X) \cdot (x \in Y)\}$$

One of the interpretations of the celebrated Boolean algebra is the area of set-theoretic operations. Another interpretation yields an idiom of Propositional Logic – the standard Propositional Logic we have been examining in this text. In fact, as you review the following characteristic Boolean Equations, reflect how corresponding equivalences in a propositional calculus would be also correct: to appreciate this, consider that complementation corresponds to what is semantically interpreted as the negation of the standard Propositional Logic, union corresponds to inclusive disjunction and intersection corresponds to conjunction. The unit of set theory corresponds to the truth value T (for true) and the null or zero element corresponds to F (for false.)

The symmetric difference of sets X and Y is defined as:

$$X \oplus Y = (X - Y) \cup (Y - X)$$

V.1 CHARACTERISTIC BOOLEAN EQUATIONS OF SET THEORY

$X \cup \Omega = \Omega$	Universal as Unit for Union
$X \cap \Omega = X$	Universal as Unit for Intersection
$X \cup \emptyset = X$	Empty as Zero-Element for Union
$X \cap \emptyset = \emptyset$	Empty as Zero-Element for Intersection
$X \cup X' = \Omega$	Excluded Middle
$X \cap X' = \emptyset$	Non-Contradiction
$X'' = X$	Involution
$X \cup X = X$	Idempotence for union
$X \cap X = X$	Idempotence for intersection
$X \cup (X \cap Y) = X$	Absorption 1

$X \cap (X \cup Y) = X$ Absorption 2
$X \cup Y = Y \cup X$ Commutation for union
$X \cap Y = Y \cap X$ Commutation for intersection
$X \cup (Y \cup Z) = (X \cup Y) \cup Z$ Association for union
$X \cap (Y \cap Z) = (X \cap Y) \cap Z$ Association for intersection
$(X \cup Y)' = (X' \cap Y')$ DeMorgan 1
$(X \cap Y)' = (X' \cup Y')$ DeMorgan 2
$X \cup (Y \cap Z) = (X \cup Y) \cap (X \cup Z)$ Distribution (unit over intersection)
$X \cap (Y \cup Z) = (X \cap Y) \cup (X \cap Z)$ Distribution (intersection over unit)
If $X \subseteq Y$, then $X \cap Y' = \emptyset$
If $X \subseteq Y$, then $X \cup Y = Y$
If $X \subseteq Y$, then $X \cap Y = X$
$\emptyset' = \Omega$
$\Omega' = \emptyset$

V.2 EXAMPLES

1. $V = \{T, F\}$

$\wp(V) = \{\{T, F\}, \{T\}, \{F\}, \emptyset\}$
$T \in V, F \in V$
$\{T\} \subset V, \{F\} \subset V, \{T, F\} \subseteq V$
$\{T\} \in \wp(V), \{F\} \in \wp(V), \{T, F\} \in \wp(V)$
$V - \{T\} = \{F\}, V - \{F\} = \{T\}$
$V \text{ x } V = V^2 = \{\langle T, T\rangle, \langle T, F\rangle, \langle F, T\rangle, \langle F, F\rangle\}$
$\{T\} \cap \{F\} = \emptyset$
$\{T\} \cup \{F\} = \{T, F\} = V$
$\mathbb{C}(V) = 2, \mathbb{C}(\wp(V)) = 4$

2. In the following extensionally defined sets, X is a subset, not a member:

$\{A, \{X\}, Y, B\}$
$\{\emptyset, \{X\}\}$

In the following sets, X is a member, not a subset:

$\{A, B, X, \{Y\}\}$
{♙, ♜, X, ∅}

In the following sets, X is both a subset and a member:

{X, {X}}

{a, {a}, b, {b}, ∅ , X, {Y}, {X}}

3. In the following set, the empty set is a member but, notice, it is also a subset because the empty set is a subset of any set (including being a subset of itself.)

   {{A}, {B}, ∅}

4. Here are examples of set-theoretic operations performed on given sets.

   Ω = { □, △, ◊, ○, ♠, ♡, ♦, ⇐, ⇒}
   A = {□, △, ◊, ○}
   B = {♠, ♡, ♦}
   A ∪ B = {□, △, ◊, ○, ♠, ♡, ♦}
   A ∩ B = ∅
   A – B = {□, △, ◊, ○}
   B – A = {♠, ♡, ♦}
   A' = { ♠, ♡, ♦, ⇐, ⇒}
   B' = {□, △, ◊, ○, ⇐, ⇒}
   A' ∩ B' = {⇐, ⇒}
   A' ∪ B' = Ω
   (A – B) ∪ (B – A) = {□, △, ◊, ○, ♠, ♡, ♦}
   (A - B) ∩ (B – A) = ∅
   A' – B' = {♠, ♡, ♦}
   (A ∩ B)' = Ω
   (A ∪ B)' = {⇐, ⇒}

5. Both the standard propositional calculus and the standard set theory can be thought of as interpretations of the Boolean algebra. We note, below, how we may map propositional connectives (in the symbolization of L, and also of IV.19) onto set-theoretic operation symbols – or, we may say, how we interpret propositional connectives by means of set-theoretic operations.

Propositional Connective	Name	Set-Theoretic Operations	Name
~*p*	Negation		Complementation
p·q	Conjunction		Intersection
p∨*q*	Inclusive Disjunction		Union
p/*q*	Sheffer Stroke	(X ∩ Y)'	
p↓*q*	Peirce Arrow	(X ∪ Y)'	
p⊃*q*	Conditional / Material Implication	X' ∪ Y = (X ∩ Y')'	
p≡*q*	Material Equivalence	(X' ∪ Y) ∩ (Y' ∪ X) = (X ∩ Y) ∪ (X' ∩ Y')	
p≢*q*	Material Disequivalence (Exclusive Disjunction)	(X ∩ Y') ∪ (X' ∩ Y)	

V.3 EXERCISES

(1) Define the following sets, given in extensional or enumeration definition, by abstraction (if possible) and by recursion. Determine the powerset and the cardinality of each set.

A = {a, b, c, d}
B = {1, 2, 3}
C = {[,]}
D = {{1}, {2}}
E = {a, 1}

(2)
In the following sets, is X a member, a subset, both, or neither? Is the

empty set a member? (The empty set is a subset of every set.) What about the set "{∅}", which is not the empty set (it is a set with one member, a singleton, whose only member is the empty set): is {∅} a member, a subset, both or neither?

a. {X, {X}, Y, {Y}, ∅}
b. {1, 2, 3, 4}
c. {∅, {Y}, {X, Y}}
d. {X, Y, Z}
e. {{∅}}
f. {{∅}, ∅}
g. {∅, {∅}, {{∅}}}

(3) Form the requested Cartesian products and powersets for the given sets.

A = {1, 2}
B = {2, 3}
C = {1, 3}

A x B, B x A, A x A, B x B, C x C, C x A, A x C, B x C, C x B
A x (B x C), B x (A x C), (A x B) x C
℘(A), ℘(B), ℘(C), ℘(A x B)

(4) Perform the requested set-theoretic operations on the given sets.

Ω = {a, b, 1, 2, △, ○}
A = {1, 2}
B = {a, b}
C = {△, ○}

A ∪ B
(A ∪ B)'
A ∩ B
(A ∩ B)'
C - (A ∩ B)
A - (B ∪ C)
B' ∩ (A ∪ C)

(A' ∩ B') ∪ (B' ∪ C')
A ∩ (B ∪ C)
C ∩ (B ∪ C)

(5) Fill out the remaining matching operations of a propositional calculus whose grammar is given as below.
PROPOSITIONAL CALCULUS: p, q, .../T, F/~ ϕ/ϕ · ψ/ ϕ ∨ ψ / ϕ ⊃ ψ / ϕ ≡ ψ /
Note the correspondences:
A': ~ p; ; A ∪ B: p ∨ q; A ∩ B: p · q; = : ≡

X ∪ Ω = Ω	(ϕ ∨ T) ≡ T
X ∩ Ω = X	(ϕ · T) ≡ ϕ
X ∪ ∅ = X	(ϕ ∨ F) ≡ ϕ
X ∩ ∅ = ∅	
X ∪ X' = Ω	
X ∩ X' = ∅	
X'' = X	
X ∪ X = X	
X ∩ X = X	
X ∪ (X ∩ Y) = X	
X ∩ (X ∪ Y) = X	
X ∪ Y = Y ∪ X	
X ∩ Y = Y ∩ X	
X ∪ (Y ∪ Z) = (X ∪ Y) ∪ Z	
X ∩ (Y ∩ Z) = (X ∩ Y) ∩ Z	
(X ∪ Y)' = (X' ∩ Y')	
(X ∩ Y)' = (X' ∪ Y')	
X ∪ (Y ∩ Z) = (X ∪ Y) ∩ (X ∪ Z)	
X ∩ (Y ∪ Z) = (X ∩ Y) ∪ (X ∩ Z)	
If X ⊆ Y, then X ∩ Y' = ∅	
If X ⊆ Y, then X ∪ Y = Y	
If X ⊆ Y, then X ∩ Y = X	
∅' = Ω	
Ω' = ∅	

Consider the following valid propositional equivalence:

$(p \supset q) \equiv (\sim p \lor q)$

What is the translation from the propositional calculus into the language of set-theoretic operations?

V.4 FUNCTIONS

A mathematical function is a relation between specified values, called inputs or arguments, and values that are defined uniquely as related to inputs, which are called outputs. A function consists in this specified relation between inputs and outputs insofar as each input has one and only one corresponding output. As the case is with sets, the inputs and outputs of functions can be anything; but they have to be specified without ambiguity and the output of each input must be uniquely defined.

We symbolize functions by using letters from the latter part of the alphabet,

{f, g, h, ...}, also allowing for subscripts from the positive integers.

Suppose, now, that f symbolizes a unary (also called a one-argument or one-place) function: this means that this function always takes exactly one input or argument and specifies one unique output for this input. If "x" refers to the input value, then "f(x)" refers to the functional output yielded by application of the function f. Other words for functions are "map" and, of course, relation.

The set of a function's inputs, which must be specified, is called the domain of the function D; the set of the specified outputs of a function is called the range (sometimes also the image or counter-domain) of the function, R.

Thus, we can define a function also as a relation between its specified domain and range.

$f: D \rightarrow R$

Having studied the set-theoretic representation of a relation, this is how we can define a unary function f:

$/f/ = \{\langle x, y\rangle / x \in D, y \in R, \text{and } y = f(x)\}$

f(x) is the output value and the ordered n-tuple, in this case ordered duplet or ordered pair $<x, f(x)>$, belongs to the set that is our function f.

This is also a complete function, meaning that the function is defined *for each value x* from the function's domain D. If there are values x from D, for which f(x) is not defined the function is called partial. (Since it is a function, the output f(x) has to be uniquely determined for each one of the values x even if the function is partial.)

Let us consider the case of the familiar algebraic addition, which is a binary function - a function that takes exactly two input or argument values to yield a determinate and unique output value. The key as to what makes this a function is that it always yields unique output values. It yields determinate values for inputs that are taken from a specified set, the domain of the function. Addition is defined for all values in its domain, so this makes it a complete rather than partial function.

What about binary functions, for example a binary function we symbolize by g?

$$g = \{<<x, y>, z> / z = g(x, y) \text{ with } x, y \in D \times D\}$$

The inputs in the case of a binary function belong to what we called the Cartesian product of the domain by itself, symbolized as "D x D" or "D^2". Thus, a binary function can be represented as:

$$g: D \times D \rightarrow R$$

Generally:

$$f^n : D^n \rightarrow R$$

An operation we can define on functions is called function composition. This is a binary operation and we will symbolize it by "$f_{o}\, g$" for two well-defined functions f and g. The functions can be of any arity. The composition operation is defined as consisting in treating the output of applying the g-function as input for subsequent application of the f-function. Thus, in the simplest case, in which both functions are monadic or unary, we have:

$f \circ g = f(g(x))$

For example,

$f(x) = x^2, g(- x):$

$f \circ g = f(g(x)) = f(- x) = (- x)^2 = x^2$

Let us consider cases in which not both functions are unary. We notice that the arity of the composing function determines, naturally, the number of required inputs for the computation.

$f(x) = - x, g(x, y) = x + y:$

$f \circ g = f(g(x, y)) = f(x + y) = - (x + y)$

$f(x, y) = x - y, g(x) = 2x:$

$f \circ g = f(g(x), g(y)) = f(2x, 2y) = 2x - 2y$

We can see that the domain and range of a composition are determined as follows. Using superscripts to the right of the function symbol to indicate arity we have:

$f^n \circ g^m = f^n(g^m(x_1, \ldots, x_m), \ldots, g^m(x_1, \ldots, x_m)) : D_{g^m} \rightarrow R_f$

V.5 EXAMPLES

(1) D = {1, 2} / R = {0}

We define unary function f.

f(1) = 0

f(2) = 0

This can be written as follows:

$1 \rightarrow 0$

$2 \rightarrow 0$

Or,

f = {<1, 0>, <2, 0>}

There are other ways of representing the function f; it is useful to familiarize ourselves with such alternative representations. For instance, see the following graph.

x	f(x)
1	0
2	0

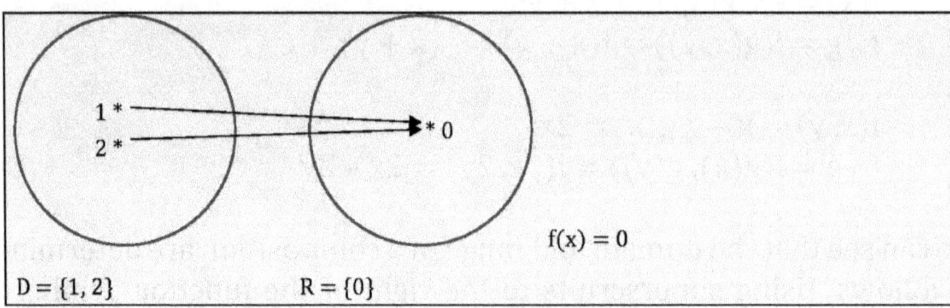

f is a function from D to R because the output value for each input value is unique. It does not matter that the output value is the same, only that it be unique. f is also a complete function.
Here is another way of representing the same function f, by means of a so-called <u>table</u>.

x	f(x)
1	0
2	0

Contrast the following function f1, defined over the same D and R.

(2) f1: $1 \rightarrow 0$
Undefined for 2

f1 is a function because every specified input assignment matches a uniquely determined output value. f1 is a partial function because it is not defined for all the members of its domain. Partial functions are rarely of interest in the kind of mathematical application we will be examining here. The truth functions of L are all complete functions - so are the mathematically definable truth functions of Propositional Logic, which include more than the ones we explicitly define in constructing PROP.

(3) The following defined assignment f2 is NOT a function because D-member 2 is assigned two values, instead of being assigned uniquely one output value, from R.

D = {1, 2}
R = {1, 0}
f2:
1 → 0
2 → 0
2 → 1

(3) Example of a binary or dyadic or two-place function. Notice that this is a complete assignment for available combinations of pairs from D and it is also a functional assignment in that a unique output is specified for each input pair.

D = {a, b} R = {a, b}
f3:
<a, a> → a
<a, b> → a
<b, a> → b
<b, b> → b

Finally, a word on Cartesian products. We saw above that the domain of a binary function is from a Cartesian product of sets to a set. The binary function always takes ordered pairs of input values and yields a determinate or unique output value for each such input pair. The members of the pair are from two sets – which may be identical with each other. The diagram below shows us how we can represent Cartesian products for given pairs of sets. We symbolize the Cartesian products of two sets X and Y by "X x Y" or, if it is the Cartesian product of a set by itself, by X x X or X^2.

Examples of Diagrams of Cartesian Products

X = {1, 2, 3} / Y = {A, B, C}	{A,	B,	C}
{1,	{<1, A>	<1, B>	<1, C>
2,	<2, A>	<2, B>	<2, C>
3}	<3, A>	<3, B>	<3, C>}

V	*V*	*V* x *V*
T	T	<T, T>
T	F	<T, F>
F	T	<F, T>
F	F	<F, F>

- *X* x *Y* = {α, β} x {β, ∅} = {<α, β>, <α, ∅>, <β, β>, <β, ∅>} ≠ *Y* x *X* = {β, ∅} x {α, β} = {<β, α>, <β, β>, <∅, α>, <∅, β>}
- W x (U x Z) = {1, 2} x ({3, 4} x {5, 6}) = {1, 2} x {<3, 5>, <3, 6>, <4, 5>, <4, 6>} = {<1, <3, 5>>, <1, <3, 6>>, <1, <4, 5>>, <1, <4, 6>>, <2, <3, 5>>, <2, <3, 6>>, <2, <4, 5>>, <2, <4, 6>>}
- ∅ x ∅ = {<x, y> / x, y ∈ ∅} = ∅

V.6 EXERCISES

(1) Which of the following monadic assignments from the given domain to the given range are not functions? Which are partial (not complete) functions?

D = {/, !, *}
R = {^, %, #}

f1:
/ → ^
! → ^
! → ^
* → ^

f2 = {</, %>, </, ^>, <!, %>, <*, %>}

f3:
/ → ^
! → %

f4 = {</, #>, <!, #>, <*, #>, <*, ^>}

(2) Which of the following binary or dyadic assignments are functions? Which are partial functions? D = {a, b} R = {1, 2, 3, 4}

g1 = {<<a, b>, 3>,<<a, a>, 1>,<<b, a>, 4>,<<b, b>, 1>}

g2 = {<<a, b>, 1>, <<b, a>, 1>}

g3 = {<<a, a>, 1>,<<a, b>, 1>,<<a, a>, 2>,<<b, a>, 3>, <<b, b>, 4>}

g4:
<a, a> → 2
<a, b> → 2
<b, a> → 2
<b, b> → 2

(3) Which of the following complete unary and binary assignments are functional?

a.
D = {T, F}
R = {T, F}

f1 = {<<T, T>, T>, <<T, F>, T>, <<F, T>, T>, <<F, F>, T>}
f2 = {<<T, T>, T>, <<T, F>, T>, <<F, T>, T>, <<F, F>, F>}
f3 = {<T, T>, <F, F>}
f4 = {<T, F>, <F, T>}
f5 = {<<T, T>, T>, <<T, F>, F>, <<T, T>, F>, <<F, T>, F>, <<F, F>, T>}

b.
D = {1, 1/2, 0}
R = {1, 1/2, 0}

f = {<1, 1/2>, <1, 0>, <1/2, 1/2>, <0, 1/2>}
g = {<1, 0>, <0, 1>}

c.
|g3| = {<<x, y>, z>/ x, y, z ∈ {1, 1/2, 0} and z = minimum(1, 1 - x + y)}

d.
g4 = {<1, 0>, <1/2, 1/2>, <0, 1>}

e.
g5:
<1, 1> → 1/2 or 1
<1, 1/2> → 1/2
<1, 0> → 0
<1/2, 1/2> → 1/2
<1/2, 0> → 1/2
<0, 1> → 0
<0, 1/2> → 0
<0, 0> → 1/2 or 0

(4) Give alternative definitions of the following functions. If not explicitly given, identify the domain and range of each function first.
f1:{1, 2, 3, 4} → {{1}, {2}, {3}, {4}}
f2 = {<1, {1}>, <2, {2}>, <3, {3}>, <4, {4}>}
f3 = <<<T, T>, T>, <<T, F>, T>, <<F, T>, F>, <<F, F>, F>>
|f4| = {x/ x = not both y and z, y ∈ {T, F}, z ∈ {T, F}};
take into account that f' = {x/ x = both y and z, y ∈ {T, F}, z ∈ {T, F}} = {<<T, T>, T>, <<T, F>, F>, <<F, T>, F>, <<F, F>, F>}, and f'' = {x / x is not y, x ∈ {T, F}, y ∈ {T, F}} = {<T, F>, <F, T>}

g1: {○, **O**, ⊙, ◎, □} → {-$100, –$50, $0, $50, $500}

g1	**-$100**	**-$50**	**$0**	**$50**	**$500**
○	x				
O		x			
⊙			x		
◎				x	
□					x

h:
D = R = {↑⥣⇡⇧↟↿1↥}
h(x) = {y/ y is an ancestor of x}
h:
1.↑ |
2. ⥣ ⇡
3. ⇧ ↟
4. ↿ 1
5. ↥

For example, <<↑, |>, <⥣>> ∈ h;
<<⥣,⇡>, <⇧>> ∈ h; etc..

(4) It has been left ambiguous whether "f(x)" denotes the function to which "f" refers or the value this function yields for specific substitutions of members of the domain set for the variable denoted by "x". This ambiguity is innocuous for most purposes but there are applications - for instance, in the enriched study of the syntactical operations of natural languages, in which disambiguation is mandated. An operation defined under the term lambda abstraction can be deployed to remove the ambiguity. The operation abstracts from the symbolization of a function (for instance, "f(x)") by isolating the variable; this is then denoted by using the Greek small letter lambda to prefix the variable: λx.f(x), or λx(f(x)). The resulting expression is defined as denoting or naming the abstract entity that is the function referred to by "f".
For specified permissible values of x, we can implement what is called lambda conversion: this consists in substituting the specified value (or values, in the case of n-ary functions) for the variable or variables. The value of the functional output can then be computed. In this way, the name of the function itself is established by lambda abstraction while the process of determination of the value of the output for specified inputs is carried out separately as lambda conversion. Examples follow:
$f(x) = x^2 + x + 1$; D = R = {0, 1, 2, ...}.
lambda-abstraction: $\lambda x.f(x) = \lambda x.(x^2 + x + 1)$

lambda-conversion for x=0: $(\lambda x.(x^2 + x + 1))(0) = 0^2 + 0 + 1 = 1$

|g| = {<<x, y>, z>/ x, y, z ∈ {1, 1/2, 0} and z = min(1, 1 – x + y)}
lambda-abstraction: λx.λy.(min(1, 1 – x + y))
lambda-conversion for x = 1/2 and y = 0:
(λx.λy.(min(1, 1 – x + y)))(1/2, 0) = min(1, 1 – 1/2 + 0) = min(1, 1/2) = 1/2

Provide the lambda-abstraction steps and the lambda-conversion steps for the indicated input values of the functions indicated below:
a. g(x, y) = max(x, y); D = R = {1, 1/2, 0}.
Conversions for all <x, y> ∈ DxD = {<1, 1>, <1, 1/2>, <1, 0>, <1/2, 1>, <1/2, 1/2>, <1/2, 0>, <0, 1>, <0, 1/2>, <0, 0>}.
b. $z = x^2 + y^2$; D = R = {0, 1, 2, ...}.
Conversions for: <x, y> ∈ {0, 1} x {1, 2}.
c.

f(x, y)	1	1/3	2/3	0
1	1	1/3	2/3	0
1/3	1/3	1/3	1/3	0
2/3	2/3	1/3	2/3	0
0	0	0	0	0

Do you detect a pattern that permits first writing the function perspicuously as a relation between the x and y variables? (Examples: f(x, y) = max(x, y); f(x, y) = min(x, y); f(x, y) = 1 – x – y; etc..)

(5) Compose the following pairs of functions to obtain $f_o g$. Determine domain and range sets for each composition operation.

f, g: {0, 1, 2, ...} → {0, 1, 2, ...}; f(x) = $\{x^{1/2}\}$; g(x) = x^2
f(x, y) = {<x, y> / x is the father of y}; g(x) = {x / x is a student in a specified class}

f: $\{T, F\}^2$ → {T, F}; f(x, y) = {<<T, T>, T>, <<T, F>, F>, <<F, T>, F>, <<F, F>, F>}
g: {T, F} → {T, F}; g(x) = {<T, F>, <F, F>}

V.7 MORE ON RELATIONS

Even though we do no aspire to be providing anything more than elementary set theory, we still have to learn certain concepts that are widely used in the study of Predicate Logic. A relation can be defined set-theoretically in terms of the ordered pairs whose members are in a specified model-theoretical extension of the relation (which can be thought of also as an n-ary predicate.) Let us have an example. Let us assume that our model has a domain – the set of all the objects we talk about. To have a proper model we also need a signature of names or individual constants labeling the objects of the domain (so that every name has exactly one object is matches) and logical predicates that are defined as subsets of the domain. (Depending on arity, the extension of an-ary logical predicate will be a subset of the nth Cartesian product of the domain.)

An illustrative model will make it easier for us to see the need for certain additional set-theoretic concepts.

D = {x/x is a positive integer} = I^+

PREDICATES = {SUCCESSOR, EVEN, ODD}

Now we write out the extensions.

EVEN = {2, 4, 6, ...}

ODD = {1, 3, 5, ...}

SUCCESSOR = {<1, 2>, <2, 3>, <3, 4>, <4, 5>, ...}

We can inspect, and seize the opportunity to reflect and consolidate understanding of, the following:

ODD $\subset$ D = I^+

EVEN $\subset$ D = I^+

SUCCESSOR $\subset$ D^2 = D x D = I^+ x I^+ = {<1, 1>, <1, 2>, <1, 3>, ..., <2, 1>, <2, 2>, <2, 3>, ..., <3, 1>, ...}

We can determine the truth value of propositional expressions in this model. No logical contradiction can be established true in any conceivable model. On the other hand, truth established in a model does not guarantee truth in other models (unless the truth is logical truth or tautology.) Expressions whose truth value is true in at least one conceivable model and false in at least one other conceivable model are logically contingent (or logically indeterminate, or, some call them, synthetic.) This is our way of formally and analytically studying relations – attaining a level of sophisticated and perspicuous management of relations, which had eluded classical logic systems like the one Aristotle put in place in ancient times.) We will soon define the concept of function and will then revisit the matter of modeling we have briefly touched upon here to detect that logical properties (including n-ary properties or relations) are basically handled as if they were expressible completely in terms of functions. It is a fascinating subject – but beyond our present scope – to examine the philosophical presuppositions for this approach to work satisfactorily (in relation to our fundamental logical expectations).

Now, let us define the <u>inverse</u> (also called converse or transpose) of a binary property and, subsequently, let us turn to the definition of certain *properties* of binary properties.
Given a relation $R = \{\langle x, y\rangle\}$, the inverse of R, symbolized as "R^{-1}", is the set of ordered pairs, $\{\langle y, x\rangle\}$. Obviously, domain and range are also reversed. The domain of R is the range of its inverse and the range of R is the domain of its inverse.

Examples can be furnished as follows:

$X1 = \{\langle a, b\rangle, \langle b, c\rangle\}$
$D(X1) = \{a, b\}$
$R(X1) = \{b, c\}$
$X1^{-1} = \{\langle b, a\rangle, \langle c, b\rangle\}$
$D(X^{-1}) = \{b, c\}$
$R(X^{-1}) = \{a, b\}$

$X2 = \{\langle \leftrightarrow, \updownarrow\rangle, \langle \rightarrow, \leftarrow\rangle, \langle \nwarrow, \nearrow\rangle, \langle \uparrow, \downarrow\rangle\}$
$D(X2) = \{\leftrightarrow, \rightarrow, \nwarrow, \uparrow\}$
$R(X2) = \{\updownarrow, \leftarrow, \nearrow, \downarrow\}$
$X2^{-1} = \{\langle \updownarrow, \leftrightarrow\rangle, \langle \leftarrow, \rightarrow\rangle, \langle \nearrow, \nwarrow\rangle, \langle \downarrow, \uparrow\rangle\}$
$D(X2^{-1}) = \{\updownarrow, \leftarrow, \nearrow, \downarrow\} = R(X2)$
$R(X2^{-1}) = \{\leftrightarrow, \rightarrow, \nwarrow, \uparrow\} = D(X2)$

$$X3 = \{<\emptyset, \emptyset> <\emptyset, \Omega> <\Omega, \Omega> <\Omega, \emptyset>\}$$
$$D(X3) = \{\emptyset, \Omega\} =$$
$$R(X3) = \{\emptyset, \Omega\}$$
$$X3^{-1} = \{<\emptyset, \emptyset>, <\Omega, \emptyset>, <\Omega, \Omega>, <\emptyset, \Omega>\}$$
$$D(X3^{-1}) = R(X3^{-1}) = D(X3) = R(X3)$$

Interestingly, logical predicates are themselves defined as relations with the ordered pairs as domain and the truth values (truth/false) as the elements of the range. Let us assume that, in a model, the extension of a binary logical predicate (or binary logical property) is as follows, over a domain D = {a, b, c}:

$$Ext(R) = \{<a, b>, <b, b>\}$$

This means that the following expressions have true as their truth value, and no other binary expressions have true as truth value (which means that they must take false as their truth value):

$$val(Rab) = val(Rbb) = T$$

We can show the domain and range of R as a logical predicate as follows:

$R: D \times D = \{<a, a>, <a, b>, <a, c>, <b, a>, <b, b>, <b, c>, <c, a>, <c, b>, <c, c>\} \rightarrow R = \{T, F\}$

We can write this out:

$|R| = \{<<a, a>, F>, <<a, b>, T>, <<a, c>, F>, <<b, a>, F>, <<b, b>, T>, <<b, c>, F>, <<c, a>, F>, <<c, b>, F>, <<c, c>, F>\}$

We also need to study certain characteristic properties of relations. We can think of them as properties of properties but, markedly, these are higher-order properties compared to the properties they characterize. To obtain an intuitive grasp of what this means, consider the proposition expressed by the sentence "red is a color:" the logical predicate we use to interpret "red" is a first-order predicate whereas the predicate corresponding to "color" has to be a second-order predicate (a predicate predicated of first-order predicates.) Higher-Order Logic falls outside our present scope.

Here we have the language of Predicate Logic (also known as First-Order Logic), which we developed in this text, to assist us, in our metalanguage, for writing out definitions of these properties. Notice how, in this metalanguage we are deploying, we also show symbolic definitions drawing on the symbols of the set theory we have been developing in this appendix. We also show a characteristic diagram of a relation that has the defined property. But it is not always possible to draw a characteristic diagram of the property: what we can do at best is diagram a relation that does or does not have the defined property. We confine our attention to binary relations.

* Reflexivity A relation R is reflexive if and only if every item has this relation with respect to itself.

∀xRxx

o ... o

↺ ... ↺

Mark that any other relationships may or may not obtain among the elements.

** Symmetry A relation R is symmetric or have the symmetry-property if and only if any two of its related members, x and y, are also related by means of the converse relation (as we defined it above.)

∀x∀y(Rxy ≡ Ryx)

o ⇆ o ... o ⇆ o

*** Transitivity A relation R is symmetric if and only if for any three of its members x, y, and z, if x and y are R-related and y and z are also R-related, then x and z must also be R-related.

$\forall x \forall y \forall z((Rxy \cdot Ryz) \supset Rxz)$

o → o ... o → o ⟶

An important triple of properties for a relation to have are Reflexivity, Symmetry and Transitivity. Such a relation R, defined over a set or domain X, is called an Equivalence Relation. It is provable that an Equivalence Relation partitions – as we say – the given domain, over which the relational predicate is defined. A partition of a set X is a set of sets – thus, a subset of the powerset of X – such that its members are mutually disjoint and their union yields X. Two sets are mutually disjoint if and only if they have no members in common or – which is the same – their intersection is the empty set. If we have a partition of a set, then for any pair <x, y> of members of X, it can be decided systematically whether <x, y> is or is not a member of the equivalence relation which partitions the set. It is also the case that, for any member S of the powerset of the set X and any partition of X $\mathbb{P}(X)$, it is decidable whether S is a member of $\mathbb{P}(X)$ or not.

We can show that any equivalence relation $\mathcal{E}(X)$, defined on a set X, defines a partition of X; and also that, given a partition of a set X, this partition defines a relation on X, which is reflexive and symmetric and transitive. We will use an example to drive home the point, leaving proof as an exercise.

There is no meaningful way for defining partitions of the empty set.

X = {a, b}
ℙ(X) = {{a}, {b}}
This is a partition because the two members, which are also members of the powerset of X, are mutually disjoint and their union is the set X (so, they can be called mutually exhaustive, although this terminology is unusual in this context).
℘(X) = {{a, b}, {a}, {b}, ∅}
{a} ∈ ℘(X)
{b} ∈ ℘(X)
It is trivial to define the equivalence relation that gives rise to this partition, Ɛ(ℙ(X)). We know, by definition, that this is a relation that is reflexive, symmetric and transitive. Recalling we can define binary relations in set-theoretic terms – as extensions – we have:
Ɛ(ℙ(X)) = Ɛ({{a}, {b}}) = {<a, a>, <b, b>}
This relation is reflexive, obviously; it is also (vacuously) symmetric and transitive. For example: it is true that, if <a, b> is a member of Ɛ, then <b, a> is also a member of Ɛ: the antecedent is false, thus making the conditional or implicative statement true. Similar considerations can be applied to show transitivity.

Let us now consider another example going in the opposite direction. Given an equivalence relation on a set Y, Ɛ(Y), we will define the characteristic partition ℙ(Ɛ(Y)).
Y = {a, b, c, d, e}
Ɛ(Y) = {<a, a>, <a, b>, <b, a>, <b, b>, <c, c>, <d, d>, <e, e>, <d, e>, <e, d>}
This is an equivalence relation: it can be ascertained that it is reflexive (<a, a> ∈ Ɛ, and so on); it is symmetric (given that <a, b> ∈ Ɛ, then <b, a> ∈ Ɛ, and so on); and it is also transitive (given that <a, b> ∈ Ɛ and <b, a> ∈ Ɛ, then <a, a> ∈ Ɛ, and so on.)
This relation defines the following partition:
ℙ(Ɛ(Y)) = {{a, b}, {c}, {d, e}}

V.8 EXAMPLES

1. We can seize the opportunity to show how we can study predicate logic by using set-theoretic means. The machinery of Set Theory is evidently put to use in the formalism that gives us standard predicate logic. This means that our ontology – the kinds of things we need to consider as available for our formal purposes – stretches rather precariously to include the abstract objects we have defined as sets. The good news, from the point of view of those who have philosophic motivations to reject a profligate ontology, is that we do not need any other abstract entities besides those of Set Theory; although the discovery of foundational paradoxes threatened to undermine this theory, our limited purposes survive any such assault. To grasp what we avoid, let us think of such famous traditional philosophic conceits as those of Platonic Forms. "Socrates is wise" is true, if it is, if and only if the individual named Socrates partakes of, or reflects the Form "wisdom' – or this Form inheres in the person. This is a deeply problematic way of speaking, extending a proliferation of spooky entities and running into notorious problems when the various details of the semantics are to be managed. Instead, we have sets as our tolerated non-concrete objects. It is true, however, that our theory is extensional and its application enforces an extensionalization: this means that, as should be obvious, we define our sets by means of their members strictly rather than intensionally or by means of notional criteria for inclusion. Thus, for instance, if it turns out that the members of "boy scouts" are exactly the same as the members of "Honors students," we must pronounce the two sets identical – but the notions are clearly distinct. We will not dwell on this subject in our present text.

Let us show how a set-theoretical setup may handle semantic problems – translated into its symbolic language. We can even translate from sentences of English. We present this briefly, by means of examples, so as to highlight the role played by set-theoretic means in modern predicate logic. Notice how we translate universal and existential, and other, statements.

(S1) All students are in class.
(S2) John is a student.
(S3) John is in class.

Tr(John) = j

SETS: S = {x/x is a student}

C = {x/x is in class}

Tr(S1): $S \cap C' = \emptyset$ [the intersection of S and the complement of C is the empty set: There are no students who are not in class.]

Tr(S2): $j \in S$

Tr(S3): $j \in C$

We can show that S3 follows validly from the other two propositions as premises:

1. Assume that (S1) and (S2) are true but (S3) is false.
2. If (S3) is false, we have: $j \notin C$; therefore, $j \in C'$.
   [by the definition of complement]
3. From (S2) we have: $j \in S$.
4. From 3 and 2: $j \in S \cap C'$
5. From 4: $S \cap C' \neq \emptyset$
   [since John is a member, this intersection set cannot be the empty set]
6. From (S1): $S \cap C' = \emptyset$
7. From 5 and 6: we have a contradiction!

Thus, by Indirect Proof, we have shown our 1, the assumption we made to be false: this means that it is possible that (S1) and (S2) are true and (S3) is false: hence, we have proven validity of the given argument.

V.9 EXERCISES

(1) Draw on the definitions in this chapter, to prove the following:

a. $F' \cup G = (F \cap G')'$
b. $F'' = F$
c. $F \cup (G \cap H) = (F \cup G) \cap (F \cup H)$
d. If $F - G = \Omega$, then either $G = F$ or $G = \emptyset$.
e. $F \oplus G = (F \cup G) - (F \cap G)$
f. If $F \subset G$, then $F \cap G \neq \emptyset$
g. $F \oplus \emptyset = F$
h. If R is Reflexive, then it has to be Serial. (Definition: A relation is serial if and only if for every domain-member x, there is some domain-member y, such that x is R-related to y.)
i. The inverse of the inverse of a relation is the relation itself.
j. The inverse of a Symmetrical relation is also symmetrical.
k. If a relation is Reflexive, Symmetrical and Transitive, - and is also called an Equivalence Relation – then it partitions its Domain: this means that, for every pair of members of the domain, x and y, either <x, y> is a member of R or it is not. Show that every Equivalence relation indeed effects a partition of its domain. Also show that there is always an Equivalence relation that can be defined by a given partition of a set.

(2) How can we approach implication in set-theoretic terms, considering that we have not defined an implicational operation explicitly? [Clue: recall the propositional rule of natural deduction, which we have called Conditional Exchange or Implication. Or, the way we can define implication from the set {~, V}, bearing in mind that the

set-theoretic analogue of the negation symbol is the complementation symbol and the set-theoretic analogue of the inclusive disjunction symbol is the union symbol.]

(3) By enlisting the formal notational machinery of PRED, we can enhance our metalanguage with PRED symbols and provide definitions of set-theoretical concepts by deploying such symbols. Examples are given below. [iff: if and only if.]

$$X \cup Y = Z \text{ iff } \forall x(x \in Z \equiv (x \in X \vee x \in Y))$$
$$X \cap Y = Z \text{ iff } \forall x(x \in Z \equiv (x \in X \cdot x \in Y))$$
$$X' = Y \text{ iff } \forall x(x \in Y \equiv x \notin X)$$
$$X - Y = Z \text{ iff } \forall x(x \in Z \equiv (x \in X \cdot x \notin Y))$$
$$\Omega = \{x \,/\, x = x\}$$
$$\emptyset = \{x \,/\, x \neq x\}$$
$$X \subseteq Y \text{ iff } \forall x(x \in X \supset x \in Y)$$
$$X \subset Y \text{ iff } \forall x \exists y((x \in X \supset x \in Y) \cdot (y \in Y \cdot y \notin X))$$

How should we use our reinforced metalinguistic notation to provide notional definitions of the following?

a. $F \oplus G = H$ iff_____
b. $X = Y$ iff ________
c. $(F \cup G)' = H$ iff_____
d. $F \cup (G \cap H) = I$ iff_____
e. $F \in$ SINGLETONS iff______ (A set is called Singleton if it has exactly one member.)
f. R is Symmetrical iff_______
g. $X \subseteq \emptyset$ iff______

(4) Given the following sets and equivalence relations defined on those sets, in each case construct the partition of the set defined by the given relation. Given the sets and partitions, define, in each case, the equivalence relation that gives rise to the partition.

X = {⚲, ●, □, ■}
$\mathcal{E}_1$(X) = {<■, ■>, <□, □>, <●, ●>, <⚲, ⚲>, <■, □>, <□, ⚲>, <□, ■>, <⚲, □>, <■, ⚲>, <⚲, ■>}

$\mathcal{E}_2(X)$ = {<⧮, ⧮>, <⧯, ⧯>, <⧬, ⧬>, <⧭, ⧭>}
$\mathbb{P}_1(X)$ = {{⧯}, {⧮, ⧬, ⧭}}
$\mathbb{P}_2(X)$ = {{⧯, ⧮}, {⧭, ⧬}}

VI. APPENDIX ON MATHEMATICAL INDUCTION

♞ The branch of Logic that studies properties of formal languages is called Metalogic. We can only cast a glimpse at this area here. As an example, we may consider the property of well-formed formulas (wffs) of a formal idiom, already mentioned, by which the number of left parentheses is equal to the number of right parentheses in every wff. The convention we have added regarding omission of parentheses for formulas and subformulas that have only associative connectives does not affect this; nor does the convention about omitting outer parentheses: in such cases the number of left parentheses dispensed with is equal to the number of right parentheses that is also dispensed with.

The process by means of which this last, or any other, property of a formal language can be shown to obtain is called Mathematical Induction on the Length or Complexity or Construction of a Well-Formed Formula. This is a special case of the proof method that is called Mathematical Induction. Generally speaking, this proof method establishes that any random member n of a (possibly infinite) collection of items that are constructed on the basis of a pattern has a property F; to do this it suffices to show that:

□ the basic instance of the pattern, known as Basis Case or Basis Step item denoted by b, has the property F; thus, F(b);

□ next, it is assumed or posited that any random item in the collection, denoted by n, has the property F; thus, a posited assumption is that

F(n) where $b < n$ in the structured order imposed on the items; this is called the Induction Hypothesis;

⚂ under the assumption that F(n), it must be shown, if possible, that F(n+1) where "n+1" denotes the items in the collection that is structurally posited as the exactly next item to the one denoted by "n" (about which the assumption in 2 was entered); this is called the Induction Step.

Once the last step has been completed, it is understood that it has been demonstrated that any random item in the collection, denoted by k and such that $b<k$, also has the property F, or that it is the case that F (k). In other words, the proof has established that any item whatsoever in the structured pattern has the property F. This is the case even if the series of items stretches to infinity.

The liberalized version we provided in the preceding paragraphs can be restricted to natural numbers. Then, showing that the number denoted by the numeral “0” has the characteristic property as the step; and, assuming that the number denoted by the numeral “n” has the property it must be shown that the number denoted by the numeral “n+1” has the property. This completes the demonstration. Returning to properties of well-formed formulas and other constructed items, we can justify application of the method by establishing a systematic parallel between the numerical method and the formal constructed elements. There have been instances in the bibliography, in which the basis case that corresponds to the numeral 0 was misrepresented; caution is needed in this respect. The catch always is that this method applies to recursively constructed entities. We will examine examples.

✇ For instance, we can show that every positive integer has a successor that is also a positive integer by use of this method. The basis case is the first number of the set, which is denoted by 1. By definition, the successor of a number denoted by "n" is the one denoted by "n + 1." The number denoted by "1 + 1", denoted by "2", is in the set of positive integers. Thus, the basis case (b = 1) has the property of having a successor. Assuming that some random - any - positive integer number denoted by "n" has a successor, by definition denoted by "n + 1", we have to prove that the number denoted by "n + 1" also has a successor. Of course, "n + 1 + 1" is equal to "n + 2" (hence, the two denote the same number.) The number denoted by "n + 2" is the successor of the number denoted by "n + 1" by definition; and this number is itself a member of the positive integers set. Thus, having assumed that the number "n" has the property, we have shown that the number "n + 1" has the same property.

Overall, we have shown in this way that any number that is a member of the set of positive integers has the property of having a successor that is itself a member of the positive integers.

⊙ This method works for numerical series that are constructed on the basis of a pattern but it is remarkable and eminently beneficial in metalogical applications given that it is operational in any case in which there is a specified recipe for the formal construction of objects according to an iterated pattern. In a Propositional Logic idiom like L, the grammar directs the user of the formal language to apply recursively specific construction rules. We have already studied how to carry out such constructions. This makes it possible to apply the method of Mathematical Induction toward showing that grammatically correct or well-formed formulas of PROP have certain properties. This is the proof procedure that is known as Mathematical Induction (or simply Induction) on the Length or Complexity of Formulas (or as Mathematical Induction on the Number of

Connectives). Based on the proof process steps in the broader Mathematical Induction method we introduced above, we have the following steps for a general target in the study of properties of well-formed formulas of a language like L:

(1) We have to show that the property holds for individual propositional letters - this is the Basis Step; this corresponds to the 0-case of numerical-series inductions.

(2, 3) Then, (the Induction Hypothesis) assuming the property to hold for any two random well-formed formulas "φ" and "ψ" we have to proceed to the Induction Step, by which we have to show:

a. that it also holds for "~ φ" (which is the same as showing for "~ ψ" also since these letters represent random formulas);

b. that the property holds and for "φ ^ ψ", where "^" is the symbol for any binary connective in PROP.

Thus, the numerical induction for the n-case corresponds to well-formed formulas of which we may think that they contain n number of connectives in them with the Induction Step showing the case for n+1 connectives. It is obvious that – something that happens often in induction proofs – the process of checking all individual cases is tedious but it is usually straightforward.

We can see how the Induction on the Length or Complexity of the well-formed Formula is a special type of Mathematical Induction. The pattern that is legislated formally by the grammar of a language like PROP dictates construction by means of a recursive - iterated and specified process. The basis case for the application of the construction is the atomic propositional variable or atom-variable, "p". The grammatical rules compel application of the tilde symbol before "p" to yield what is defined in the grammar as correctly constructed formula "~ p". For any well-formed or grammatically correct formula "φ", application of the tilde before "φ" yields "~ φ" which is also specified in the grammar as well-formed. For any binary connective symbol "^", the recursive grammar specifies that any two well-formed formulas (wffs) "φ" and "ψ" may be combined in infix notation (with the binary connective symbol in between the combined wffs) to yield wff "φ ^ ψ". The basis for the construction is "p" and whatever properties it exemplifies are retained when the construction yields a complex formula based on iterated uses of atom-variables in accordance with the rules. For instance, given that "p" has no connectives, "~ p" has the connectives of "p" plus the added unary connective "~" and, so, has one connective. Any complex wff has a number of unary connectives which depends straightforwardly and exclusively on iterations of the patterned process of construction mandated by the grammar of PROP. Assuming any random case of a wff, "φ", which is not atomic, to have number of unary connectives n, then "~ φ" has n + 1 unary connectives. If combined with another wff, "ψ", the resultant wff has a number of unary connective that is the sum of n and whatever number of unary connectives "ψ" has. The function of the number of unary connectives of a wff is thus definable:

0 if φ is atomic

UC(φ) = UC(ψ) + 1 if φ = ~ ψ

UC(x) + UC(ψ) if φ = x ^ ψ, where "^" denotes any binary connective.

No other property has this characteristic function. This shows us, again, that mathematical induction allows us to track instantiation of a unique property across all grammatically correct formulas (wffs) of the formal system PROP.

VI.1 EXAMPLES

1. We will first apply the method of Mathematical Induction on the Length of a Formula to demonstrate that every well-formed formula of PROP has equal numbers of left and right parentheses.

a. Basis Step: p has no parentheses, as dictated by the grammar of PROP. So, the numbers of left parentheses (l) and of right parentheses (r),

$l(p) = r(p) = 0$

b. Induction Step: Assume, as Induction Hypothesis, for any wffs "φ" and "ψ" of Lthat

$$l(\varphi) = r(\varphi) \text{ and } l(\psi) = r(\psi).$$

"~": No added parentheses are needed when [~] is placed in front of a well-formed formula "φ". Same proof for "ψ".

$$l(\sim \varphi) = l(\varphi) + 0 = l(\varphi)$$

$$r(\sim \varphi) = r(\varphi) + 0 = r(\varphi)$$

But, by the assumption we have made, $l(\varphi) = r(\varphi)$

Hence, $l(\sim \varphi) = r(\sim \varphi)$

"^": Now for the binary logical constants. When two complex formulas are connected by binaries,

$l(\varphi \wedge \psi] = l(\varphi) + l(\psi) + 1$

$r(\varphi \wedge \psi) = r(\varphi) + r(\psi) + 1$

This is the general case. We have allowed for omitting outer parentheses and parentheses around formulas and parts of formulas which have only associative connectives ({V, ·, ≡}); this does not affect the outcome since this elimination removes equal numbers of left and right parentheses.

So, for any binary-connective symbol "^":

$l(\varphi \wedge \psi) = l(\varphi) + l(\psi) + 1$

$r(\varphi \wedge \psi) = r(\varphi) + r(\psi) + 1$

But, $l(\varphi) = r(\varphi)$ and $l(\psi) = r(\psi)$

Therefore, $l(\varphi \wedge \psi) = r(\varphi \wedge \psi)$

♝ We can show the relationship between apparently non-numerical applications of Mathematical Induction and numerical ones by means of a more perspicuous way of carrying out proof of the property of well-formed formulas (wffs) we just considered. This is the property by which each wff has merely a finite number of atomic propositional variable letters.

The basis step is the case for which the smallest natural number can be assigned. The set of natural numbers is N = {0, 1, 2, ...}. Basis step corresponds to n = 1: any wff of Lmust have at least one symbol and the minimal case is the one with one propositional variable letter.

1) Basis Step: n = 1 -> wff = p -> #(atomic variable letters) = 1 ∈ FINITE

2) The Induction Hypothesis: for all $k \in N$, such that $1 < k < n \in N$,

#(atomic variable letters in wff) ∈ N

The key here is to assign natural numbers to wffs so that those numbers track the iterated and regulated construction of wffs from simpler wffs as stipulated by the grammar of PROP. This numeration is available as we record the length of a wff, defended as the number of all the logical symbols of a wff (all symbols except the auxiliary parentheses.) For the basis case, where n = 1 (the length of the wff is indeed 1), we have checked and determined that the property (finite number of atomic letters) obtains. The Induction Hypothesis has posited, in accordance with the method of Mathematical Induction, that the property also obtains for all lengths between 1 and some length number, n (any random such number). We will then have to demonstrate that a wff of n-length also displays the property of finite number of atomic letters. Notice that we are following a stricter method of Mathematical Induction - called usually Complete Mathematical Induction - which rests on the inductive assumption that all members of the patterned series up to some constructive step display the property.

3) A wff of length n can have as major function symbol one of {~, ·, ∨, ⊃, ≡, ≢}, as we examine the case of an idiom PROP ∪ {≢}, the formal language PROP enhanced with the symbol and conventions for the disequivalence sign. All cases must be examined.

Assume that the n-length wff has the tilde as major operator or function symbol - this is the case of negation. Call this formula ψ = ~ φ. Notice that the formula denoted by "φ" has length that is less than n. Accordingly, it is covered by the induction hypothesis: it has a finite number of atomic letter symbols. But the wff denoted by "ψ" has the same number of atomic letter symbols as (the one denoted by) "φ". Therefore, ψ (henceforth omitting the circumlocution and the brackets but referring to the wffs still) has a finite number of atomic letter symbols.

Next case is of an n-length wff $x = \varphi \wedge \psi$, where "^" denotes any one of the binary operator symbols of PROP. Notice again that both φ and ψ have lengths which are, each, less than n. They are, therefore, covered by the induction hypothesis and have, each, a finite number of atomic symbols. The wff x has as number of symbols, #(atomic symbols in x), the sum of the numbers of atomic symbols in the two component wffs, #(atomic letters in φ) + #(atomic letters in ψ). Thus, the number of atomic letter symbols in x is also finite.

2. Next, we apply mathematical induction on the complexity of well-formed formulas to prove that any wff of PROP has a finite number of propositional letters.

a. Basis step:

p has a finite number of letters (= 1). So,

$\#(p) = 1 \in$ FINITE (We add "$\in$" as a METALANGUAGE symbol for "it belongs to set...")

b. Induction Step:

Assume that any two wffs of L"φ" and "ψ" have each a finite number of propositional letters, respectively m and n.

$\#(\varphi) = m \in$ FINITE

$\#(\psi) = n \in$ FINITE

"~": $\#(\sim \varphi) = m \in$ FINITE

(The same is the case for "~ ψ" - but this need not be mentioned at all since what matters is that "ψ" is a random wff; indeed, for the assumption given above, $\#(\sim \psi) = n \in$ FINITE)

"^": For any one of our binary connective symbols ({·, ∨, ⊃, ≡, ≢}),

$\#(\varphi \wedge \psi) = \#(\varphi) + \#(\psi) = m + n$

Since $\#(\varphi) = m \in$ FINITE and $\#(\psi) = n \in$ FINITE,

$\#(\varphi \wedge \psi) = m + n \in$ FINITE

3. Proceeding conversely, it is possible to determine *what property of a well-formed formula of Lis denoted by a given function.* In this case we are given the function and determine the property whereas in what we did in the preceding exercises was to construct this function in order to demonstrate that formulas of L have the indicated property. Accordingly, the definition of the function, which we will examine now, has to be presented for the basis case and for the n and (n+1)th cases for the induction step.

For example: Consider the following function and determine what property of well-formed formulas of Lthis function denotes.

$@(p) = 0$

$@ (\sim \varphi) = @ \varphi$

$@ (\varphi \wedge \psi) = @(\varphi) + @(\psi) + 1$ (for any binary connective of PROP, denoted by "^")

This function denotes the number of binary connectives of a well-formed formula of PROP.

VI.2 EXERCISES

(1) Use mathematical induction on the length or complexity of a well-formed formula to demonstrate the following:

> Every wff of PROP has the property of compositionality defined as strict functional dependence of the meaning of the compound on the meanings of the components. Understand that by "meaning" we are referring to logical meaning or referent of the proposition denoted by a symbol: this a truth value from the set {T, F}.
>
> Call the total number of symbols, omitting auxiliary parentheses, a well-formed formula length of the formula and symbolize by PROP. Define as weight of a well-formed formula, symbolized by w, the sum of the individual weights of all the unary and binary function symbols in the formula, where the individual weights are 1 for the unary function symbol, {~}, and 2 for each of the binary function symbols. For example, the length and weight of $\varphi = (p \supset (q \equiv (\sim q \cdot t)))$ are: $l(\varphi) = 8$ and $w(\varphi) = 2 + 2 + 1 + 2 = 7$. Use mathematical induction on the length of the formula to show that the length and weight of every wff φ of Lare related as follows: $l(\varphi) = w(\varphi) + 1$.
>
> Use induction on the length of the formula to show that no well-formed formula of Lcan terminate in a tilde.
>
> Use induction on the length of the formula to show that the members of the set $\{\sim, \cdot, \vee, \supset, \equiv, \not\equiv\}$ are the only connective symbols that can appear in a well-formed formula of PROP.

(2) What properties of well-formed formulas of PROP are characterized by the following functions? "^" is a metalinguistic symbol for any connective of the grammar of PROP. Take into

consideration that outer parentheses are omitted by special convenient permission in the grammatically correct constructions of PROP.

$f_1(p) = 1; f_1(\sim \varphi) = f_1(\varphi) + 1; f_1(\varphi \wedge \psi) = f_1(\varphi) + f_1(\psi) + 1.$

$f_2(p) = 0; f_2(\sim \varphi) = f_2(\varphi) + 1; f_2(\varphi \wedge \psi) = f_2(\varphi) + f_2(\psi) + 1.$

$f_3(p) = 0; f_3(\sim \varphi) = f_3(\varphi); f_3(\varphi \wedge \psi) = f_3(\varphi) + f_3(\psi).$

$f_4(p) = 0; f_4(\sim \varphi) = f_4(\varphi) + 1; f_4(\varphi \wedge \psi) = f_4(\varphi) + f_4(\psi) + 2.$

(3) Discover the function that characterizes the number of left parentheses of a wff of L when you do not omit outer parentheses. Next, consider the function that characterizes the number of binary connective symbols in a wff of PROP. Discern a relationship between the two functions. Then, prove rigorously that this relationship indeed obtains by using induction on the length or complexity of the formula.

VII. APPENDIX ON DIALOGUE GAMES FOR PROPOSITIONAL AND PREDICATE LOGIC

VII.1 DIALOGICAL LOGIC (LDLG)

Dialogical Logic was invented by Paul Lorenzen in the late 1950s' and further developed in collaboration with Kuno Lorenz (with the results encapsulated in *Dialogische Logik* (Darmstadt, 1978)); Jaakko Hintikka and others have also made seminal contributions to the technical expansion and philosophic investigation of this program. Strictly speaking, Dialogical Logic is a framework that draws on the resources of zero-sum, incomplete-information game theory to implement a variety of logical languages; this variable implementation can be accomplished by tweaking appropriate dialogical rules. In this way, dialogical approaches to the semantics of logics give a more detailed account of what happens when we shift from one logic to another: in addition to a change in the definitions of connectives, shifts from one logic to another seem to require changes that affect constraints on winning game-strategies. This may indicate that a deeper view of what constitutes a logic is broader than allowed by the standard view that characterizes logical systems at the level of their connectives or logical constants. While the traditionalist philosopher of logic may dismiss claims about a formal variety of logical systems as a mirage created by playing with alternative definitions of logical connectives, the dialogical reconstruction of different logics shows that something more is involved in the deeper constitution of a logical language. Thus, the philosophic position known as Logical Pluralism gains theoretical reinforcement by the success of logic-games in implementing familiar logical systems.

We present here a dialogical system for Propositional Logic, Ldlg. Strictly speaking, the transactions in the system are to be carried out in a metalanguage that is enhanced with appropriate symbols, which are to be specified subsequently. This enhanced metalanguage we designate Ldlg for referential convenience.

There is something natural about the dialogical logic method as a decision procedure - elegantly fulfilling an implicit promise that dates back to the Socratic approach to philosophical inquiry and also echoes the disquisitional method of Medieval *Obligationes*. The thesis laid down, to be defended by a proponent, is shown to be a validity of the logic if and only if the defender or proponent has a winning strategy against another player, the opponent who also seeks to implement an ideal winning strategy toward refuting the thesis. The game is one of incomplete information because utterance of formulas is not assessed on the basis of justification or supported claims about the truth of the formulas that are uttered. This means also that winning strategy becomes the key notion and this notion replaces validity or logical consequence. It is not truth-preservation that is considered to be constitutive of validity but availability of a winning strategy to the idealized proponent of a thesis, who implements the best available strategy for winning (as does the opponent.) The dialogical approach is not fashioned after the model-approach to semantics, either: there is no burden to flesh out claims made by the players by means of appropriate objects - as is done with models. Rather than show that semantic properties like truth are fixed across all possible models, the dialogical approach makes checks of validity and tautologousness depend on winning a game that is winnable in accordance with specified rules. Such rules are set not only for the connectives but also for the structure of the game processes or moves.

The technical implications of the availability of the dialogical framework for logics should not be underestimated. Dialogical approaches can be adapted, by suitable manipulations of their rules, to yield systems whose semantic interpretations are exactly matching not only standard Propositional Logic but also, with different rules, non-classical propositional systems like the one known as Intuitionistic Logic. What is philosophically remarkable is the following result: We obtain the correct systems for the various logics by manipulating not the rules for the logical connectives but the other kind of rules mentioned above, which are known as structural rules. The orthodox view is that a logical language is determined ultimately by characteristic definitions of its logical connectives (also called logical constants or, for extensional systems, truth functions.) Structural rules, on the other hand, appear to capture aspects that correspond to pragmatics - matters of linguistic usage rather than the formal elements, associated with the definitions of the connectives, which have been taken as sufficient for determining what a logic is.

VII.1 STRUCTURAL RULES

In the construction of dialogical systems, the right variation in the structural rules gives us a corresponding system that can be semantically interpreted as Intuitionistic rather than as classical propositional. We will not pursue this alternative construction here but the fact is worth mentioning. When Harvard logician and philosopher W. V. O. Quine ridiculed authors of non-classical logical languages, he scoffed that "they are changing the subject" when they define new, instead of the classical, logical connectives. This redefinition is inevitable when there is a shift from one logical language to another. Assuming that non-classical logicians are sufficiently competent to realize that the connectives of non-standard systems have different definitions, Quine's deeper complaint must be that the reasons that may dictate a shift in the meaning of a logical connective are pragmatic - having to do with linguistic usage - and, as such, they are not relevant to the purely formalist concerns of constructing systems in deductive logic. Quine was wrong to think that there is something lacking, in formalist terms, when it comes to non-classical systems; but he is also shown to be wrong in a deeper sense because pragmatics, mimicked by enforcing variations in structural rules of games, result in constructions of logical systems that are formally different from each other. As we will see, structural rules play the role that usage rules play in languages and, in this respect, they are unlike the rules that regulate moves for introducing and removing the logical connectives. It turns out, then, that logical formalism may legitimately extend beyond the settings for the meanings of connectives and, so, has a broader scope than it was initially thought.

Dialogical Logic has a pragmatic flavor to it from the very beginning because it includes structural or game-regulating rules in addition to the rules for the logical connectives. Here we will construct a system we call PROPdlg, which, in addition to rules that govern use of the system's connectives (connectives rules), also has structural rules, which specify what is mandated, and what is not permissible, with respect to initiating and terminating moves in the kind of game that is constitutive of dialogical setup for classical Propositional Logic. The connectives rules, also called Local Rules, reflect the definitions of the propositional connectives (or logical constants, i.e. the truth functions of standard Propositional Logic.) Such rules are independent of conversational content or contextual considerations; this is what ought to be expected given the experience-transcending character of deductive reasoning. At the same time, the local or connectives rules are so arranged in Dialogical Logic as to propel steps in what seems like an ongoing exchange between two players, one of whom defends, while the other attacks, a thesis. Both players are compelled both to attack each other and to defend against attacks launched against them in observation of the ideal best strategy available to them. The Proponent is the players who initiates the game by laying down or asserting the thesis (well-formed formula of the system) that is to be proven or disproven. The effective reliance of the dialogical setup on structural rules is remarkable and has considerable philosophic implications possibly hinting at a deeper relationship between the content-independent elements of formalisms and the suitable applications of such formalisms. A narrowly formalist view of logic, insisting on the inviolability of the demarcation of the formal aspects of Logic from linguistic usage, is hard pressed to explain how it is possible that structural rules – which are contextually and contentfully driven rather than depend on the formal definitions of the connectives of the language – can have an impact in characterizing the logic itself. Characterization here means this: a system like Ldlg can be interpreted semantically and then it matches L, which, as we have seen in this text, is an idiom of the standard Propositional Logic. Thus, Ldlg has a distinguishing interpretation that matches PROP. With appropriate and characteristic adjustments to the structural rules, and as we obtain some other system, for instance Ldlgint, we now have a system that lends itself to an interpretation as an idiom of Intuitionistic Logic. More on this remarkable result follows.

The structural rules distinguish two different kinds of players and divide labor between them, lay out two broad types of movement (characterized as attack and defense, following an initial step of laying down the thesis to be defended), and specify what kinds of movements are permissible to undertake in order to attack (or challenge) or defend against previous attacks and when such moves may be played. A refinement in the structural rules, itself being such a rule, allows the proponent to repeat defenses against the same prior move that has already been defended once and, generally, to defer immediate defense and to defend against earlier attacks rather than only against the latest attack. This is crucial because it is this tweaking in the structural rules that turns out to be the differentiating factor between the standard Propositional Logic and the non-classical logic (a sub-logic of the classical Propositional Logic) known as Intuitionistic Logic. When the defense against earlier attacks, even if later attacks can be parried, is allowed, the system yields exactly the theses whose interpretations constitute classical logic; when this delayed attack is not permitted, we obtain Intuitionistic Logic. Addition of the delayed-defense rule extends logic from Intuitionistic to Propositional: all the tautologies of Intuitionistic Logic are still provable but new tautologies become provable under the extension. This is as it ought to be given that Intuitionistic Propositional Logic is a sub-logic of the Classical.

The basic setup of dialogical logic has a proponent and an opponent. The proponent lays down what is to be proven as a thesis of the logical system; if, and only if, the proponent wins do we have the proposed formula as a thesis of the system. If the opponent wins, on the other hand, the proposed or propounded formula is refuted and does not obtain the status of being a thesis of the system. To check whether consequence obtains in this way, by constructing a dialogue for a thesis, we take advantage of the fact that the Theorem of Deduction is provable in the case of Propositional Logic. For well-formed formulas (wffs) φ1, ..., φn, and ψ, and the connectives interpreted semantically by the conjunction and material implication of PROP (which we can symbolize here too by "·" and "⊃"), we have:

(Deduction Theorem) $\varphi 1, ..., \varphi n \vdash \psi$ if and only if $\vdash (\varphi 1 \cdot ... \cdot \varphi n) \supset \psi$

In other words, premises and conclusion (in the semantic interpretation) are related by logical consequence if and only if the implicative statement with the conjunction of the premises as antecedent and the conclusion as consequent is a logical truth of the semantic system. Accordingly, in dialogical logic, we can test whether consequence obtains by having the proponent lay down the implicative statement formed as an implication with the conjunction of the premises as antecedent and the conclusion as consequent. This is sometimes called a sequent. The proponent lays down the sequent, as the opening move or gambit of the decision process. Then, the opponent attempts to refute it; if the opponent fails in accordance with the rules of the system, the implicative statement is shown to be a thesis of the dialogical system: this further means that the premise(s) constituting the conjuncts in the antecedent of the sequent have as logical consequence (entail) as conclusion the consequent of the thesis. Failure is defined as unavailability of any further move (either defensive or attacking move). Failure of one player automatically means that the other player wins.

We will present a summation of the structural rules after we have introduced the formal notational language for Ldlg.

We construct Ldlg with the same symbols as we used for the initial Propositional Logic idiom we constructed under the name PROP. There is no risk of ambiguity but it ought to be remembered that the abstract objects over which Ldlg is constructed are not the same as those of PROP. For atomic letters, we may draw from the denumerable stock {p, q, r, ...}, with subscripts allowed, and for well-formed formulas from the denumerable set {φ, ψ, ...}, also with subscripts allowed. For connectives, we restrict ourselves to those symbolized by {~, ·, V, ⊃ , ≡}. Next, we present the connectives rules. Each connective has three rules which are defined sequentially with (x + 1) rule understood as available only after x rule has been applied. The connectives rules themselves are structurally inflected: as we will see, these rules are specified within pragmatic contexts; there is an opening rule, triggered when a formula with the connective as its main connective, is laid down; subsequently, there is a rule for attacking a connective and a rule for defending a connective after it has been attacked. Only one connective - the negation connective - lacks a rule of defense against a launched attack. Structural rules also specify when a deferred defense or a repeated attack may be allowed. For some connectives, the rules may require branching, similar to the situation in implementations of the kind of methods known as tree or tableau. Indeed, there is a close correspondence between the dialogical decision process steps and those in implementing a tree decision procedure; it is possible to match the parallel moves in the two procedures (dialogical and tree) step by step. This demystifies the mechanics behind the spectacular success of the dialogical method but it does not detract from the significance of having such an effective method as a decision procedure for propositional (and for predicate) logics both of classical and non-classical varieties. Let us introduce the formal language Ldlg.

<u>* The Formal Language Ldlg</u>

Propositional Variables or Atoms: {p, q, ...}, denumerable stock with subscripts allowed.

wff (well-formed formulas, not necessarily atomic): {φ, ψ, ...}, denumerable stock with subscripts allowed.

Connectives = {~, ∨, ·, ⊃}

Material equivalence is defined derivatively as:

$(p \equiv q) =_{df} ((p \supset q) \cdot (q \supset p))$

Formation rules for the connectives as in PROP. The grammar of Ldlg requires additional symbolic expressions.

♛ A wff of Ldlg has the form:

<PlayerType, SignatureMark, φ, MoveType, ConnectiveRuleUsed, ConnectiveUttered>

Logical Consequence Symbols: {⊢, ⊬}

We also use numbers from the positive integers for the lines.

♚ Player Types = {P, O}

P: proponent or defender

O: opponent or challenger

♜ Signature Marks = {!, ?}

"!" signs or prefixes a wff (well-formed formula) that is defended but there are exceptions as the connective rules, below, will specify.

"?" signs or precedes an attack on a wff that has already been laid down. It may be an attack against a move made earlier and not yet challenged.

♞ Move Types = {Th, Aj, Dk}

Th: laying down the thesis to be proven or disproven (opening move or gambit)

A: attack or challenge

D: defense

Attacks (or challenges) and defenses are followed by numbers indicating, for attacks, the rank of the move (e.g., "A1" means first attack by opponent in the game), and, for defenses, the corresponding rank of the attack against which the defense is offered (e.g., "D1" means defending against the first attack.) Attacks and defenses are specific to the connectives and are regulated in accordance with the connective rules.

VII.2 CONNECTIVE RULES = {~1, ~2, ·1, ·2, ·3, V1, V2, V3, ⊃1, ⊃2, ⊃3}
Connective Rules are used for denoting a move (attack or defense) and for laying down or uttering the connective (actually, as major connective of the formula laid down.) The first rule for each connective simply permits laying down a wff with this connective as its major operator. For instance, "~1" indicates that the player is making a move that initiates or lays down a negated formula; this formula now may be attacked (with such an attack, appropriate to negation, indicated then by "~2.") Notice that "~3" is missing from the set of connectives rules symbols above, which, indeed, means that there is no defense available for an attack made against a negation.

☻ Given the grammar of Ldlg, an example of what a line may look like is:

m. O! (p ⊃ q) A1 ⊃2 ⊃1.

This reads as: "in line m, the opponent lays down the wff 'p ⊃ q' and this constitutes the first attack or challenge in the game; this is done by using the connective rule '⊃2' (hence, it must be an attack against a previously posited implicative formula, for instance "(p ⊃ q) ⊃ ~ p"); and the formula laid down in this attack is itself an implicative formula and, so, accords with connective rule ⊃1."

After we have studied the rules for the connectives, we will be able to tell that this move is indeed an attack, permitted by the rules for "⊃", in spite of the use of the signature "!" - which, as we mentioned earlier, is often but not necessarily a signature of thesis-assertion or defensive moves.

☞ Ldlg examines well-formed formulas for tautologousness. So, it renders decisions on argument validity by means of a dialogue for the corresponding implicative formula that is formed by taking all the premises conjunctively as antecedent and the conclusion as consequent: for instance, to check validity of the argument form

$\varphi, \psi \vdash x$

we construct the dialogue

$dlg((\varphi \cdot \psi) \supset x)$

In this way we have an implicative sequent to check by means available in Ldlg. As indicated above, validity essentially corresponds to availability of a winning strategy for the idealized Proponent who opens the dialogue by positing the thesis.

♅ Structural Rules

⚃ (RO) Game Opening: The dialogue (or dialogical game) begins with the proponent laying down the thesis (Th) that is to be defended. We say that the ensuing dialogue is a dialogue for the formula that has been laid down as thesis: dlg(φ).

The thesis itself, as well-formed formula of Ldlg, has a main connective *; laying the thesis down is itself in accordance with the rule *1 which governs utterance of that connective.

The opponent carries out the first attack or challenge. To do so, the opponent has to utilize the appropriate connective rule (to be presented below) depending on which connective is attacked; this, of course, would have to be the main connective in the formula that has been laid down by proponent and is now attacked by the opponent. Thus, the Opponent is acting in accordance with rule *2, while uttering some new formula in accordance with rule #1 where "#" is the symbol for the main connective of the Opponent's challenging formula.

⚃ (RG) Game Process: The game's first segment is initiated, as we saw, by the opponent attacking or challenging the thesis laid down by proponent. In Ldlg, the players are idealized and maximally competent. Indeed, what is assessed is availability of winning strategies. Empirical limitations are never stipulated. *Each player is assumed to be using the best available strategy*; so, losing is never a matter of failing to utilize an available but undetected winning strategy for the player. Compare this with the application of a procedure like the truth table: empirical cases of errors on the part of the actual user of the truth table are irrelevant.

The game continues, after the initial attack by the opponent, with either a defense against the attack that has been launched or with a new attack that must be defended subsequently. Both for attacks and defenses, the players must use the available connectives rules which, as we will see, are defined systematically to be different depending on whether the connective is attacked or defended.

The rules for the connectives are utilized in the same fashion by the two kinds of player in the game. In this sense, there is symmetry - unlike in the case of tree or tableau rules, where different connectives rules apply depending on whether formulas are asserted or negated.

A special structural rule, presented below, will specify that attacks and defenses may be postponed. This means that responses are not necessarily defensive: one may postpone a defense and attack instead as a matter of responding to an attack.

✇ (RD) Rule About Strategic Delays. It is ensured that the game does not run into infinity of moves by means of this rule which prohibits open-ended postponements or repetitions (with the exception of a repeated defense allowed for the Proponent). Delays that are not open-ended or indefinite are permissible and play a significant role in differentiating logical systems: for instance, a prohibition on a delayed defense makes possible proof of formulas that are tautologies for classical but not for intuitionistic Propositional Logic.

⚀ (RA) Availability of Atoms: The Proponent is allowed to lay down an atom (atomic or single propositional letter) only if the Opponent has already conceded it or laid it down. In other words, any atomic letter becomes available to use only after the Opponent lays it down.

◻ (RPost, RRep) Postponements and Repetitions of Attacks/Defenses: Defenses may be postponed if they are not immediately available and/or if other moves - including attacking moves - are available.

A crucial structural rule is this: the proponent is allowed to repeat an earlier defense. Without this rule we cannot prove certain tautologies of the standard propositional system. We obtain, however, exactly another system, all of whose tautologies we can prove, and this is the non-classical system known as Intuitionistic Propositional Logic.

◻ (RT) Game Termination: The game ends when one of the players has absolutely no available move to make either in attacking or in defending against an attack already made - (including rightly deferred defenses or rightly postponed attacks). When the game so ends, Proponent wins if and only if at least one atom (atomic or single propositional letter) has been laid down or uttered by both proponent and opponent. In this case, the thesis laid down for the dialogue has been shown to be a tautology or validity of Ldlg.

If the proponent has no more moves to make to defend against previous attacks, the Opponent wins if, additionally, at least one atomic variable letter has been uttered by both players. The formula that was laid down as thesis has been shown not to be a tautology of Ldlg.

No other possibilities for closure are available. It can be shown that games cannot run into infinity for Propositional Logic (see also the rule against strategic delays or open-ended postponements and repetitions above.)

◊ (Closure Clause) Nothing else is acceptable as a structural rule for Ldlg.

In summary, the structural rules for the Ldlg are as follows:

Player – signature	**wff uttered**	**Action**	**Structural Rule**
1. P!	φ	Th	RO: opening of the game; Proponent states thesis to be proven or disproven.
2. O!	ψ	A1	RG: Rules of Process for the Game. Attack or Challenge by the Opponent; the form the attack takes depends on what is the major connective in the formula laid down by the Proponent.
3. P!	...	D1/A2	RG: The Proponent may defend, if possible, by utilizing the connective rule for defenses for the major connective of φ (the thesis). Or, the Proponent may attack ψ by utilizing the appropriate attacking rule for the major connective of ψ (the formula laid down by O in A1). Or, the Proponent
4. ...	...	D_j/A_k	RPost: Defenses to previous moves may be postponed or undertaken with delay, punctuated with attacks in between. This is a strategic consideration; more broadly, each player is conceived ideally or as pursuing the best available

				strategy for winning, if possible; such strategy may include deferring or postponing defenses. There is, however, a structural rule that is needed to block infinitarian extensions of games: RD: the delay to defending cannot be open-ended in the sense that the defense has to be entered if no other better move is available; the same applies in the case of repeated defenses.
5.	**...**	...	...	RRep: Defenses may be repeated by Proponent. This applies also in the case a defense comprises offering either one of two conjoined formulas (disjuncts or conjuncts.) If this rule is dispensed with, while maintaining all the other rules, we construct exactly the weaker logical system, or sub-logic, known as Intuitionistic Propositional Logic.
6.	**...**	⊗	No moves Available	RT: Termination Rule. If either player has no available move, the other player is declared as winner.

Each connective (generically symbolized by "#") has three sequential rules associated with it. The first rule merely tracks utterance, indicating that a well-formed formula with the connective as main operator has been laid down.

The second and third rules specify, respectively, how the connective may be attacked and how it may be defended against an attack. These rules essentially define the connective in the system Ldlg.

#1. The first rule is about what counts as a laying out of the connective (actually, laying down of a wff with the connective as its main operator.)

#2. The second rule, available only after rule one has been satisfied, specifies how the connective (the wff with the connective as main operator) may be attacked or challenged.

#3. Finally, the third rule specifies how the connective, once already attacked per rule 2, may be defended.

Negation Rules
Player1: ! ~ φ ~1
Player2: ! φ ~2 A_j
[No defense for negation – no 3rd move available]

Conjunction Rules
Player1! φ · ψ ·1
Player2? · left or Player2 · ?right ·2 Aj
[attacker chooses]
Player1! φ or Player1! ψ ·3 Dj

Inclusive Disjunction Rules
Player1! φ ∨ ψ ∨1
Player2 ? ∨ ∨2 Aj
[attacker does not choose]
Player1! φ or Player1!ψ ∨3 Dj
[choice lies with the defender]

Material Conditional Rules
Player1! φ ⊃ ψ ⊃1
Player2! φ ⊃2 Aj
Player3! ψ ⊃3 Dj

In accordance with the structural rules for PROPdlg, the Defender may not produce an atom (atomic or single propositional variable) unless the Attacker has already stated it and, thus, made it available. Other crucial structural permissions and restrictions involve postponement of defense on the part of the defending player, like repetition of a defense already undertaken (when there is choice and one, but not the other, disjunct or conjunct has been produced.) How such rules are adjusted affects the species of logic for which the dialogical setup can serve as a decision procedure.

Note that the attack may use either signature from {!, ?}, depending on what the connective is. In other words, it is not always a "?" sign that signals that an attack is under way. Negations and conditionals are attacked by counter-asserting, thus using the "!" sign. In the case of a negation, the formula that is negated is being asserted as a matter of attacking while for conditionals the attacker asserts the antecedent demanding that the consequent be produced by the defender. The player who has laid down the negation is dared to defend against that counter-claim that the formula is assertable! Of course, there is no defense against this move because no arbitrary formula (including negations of formulas) carries a presumptive right to being asserted. In the case of attacking a material conditional, the challenger asserts the antecedent and dares the defender of the conditional to produce the consequent in defense. In many games, the Proponent who faces this attack cannot yet produce the consequent because this consequent is an atomic letter that has not yet been conceded or asserted by the Opponent. (Recall the structural rule that specifies that atoms become available to the Proponent to lay down only after they have been conceded by the Opponent.) Of course, if other moves are available, the defender against assertion of the antecedent (attack on the conditional) may execute one of those moves: such permissible moves may be deferred defenses or attacks against earlier utterances made by the other player.

The similarity between the connective rules for negation and for material implication is not a coincidence. Since dialogues, adjusted accordingly by means of tweaking structural rules, can yield sub-logics of the standard Propositional Logic, their treatment of negation is, in a deeper sense, to be understood as defining "$\sim p$" as "$p \supset \bot$" where "$\bot$" is the symbol denoting logical absurdity. Hence, the challenge to laying down a negation, as with the challenge to material implication, consists in laying down the antecedent of "$p \supset \bot$" and daring the other party to produce an absurdity, which, of course, cannot be ever asserted properly: it follows that there is no available defense against the challenge to negation and this can be grasped intuitively by noticing that the assertability of neither atomic propositions nor their negations can be negotiated by means of the rules of the connectives of a propositional system.

⊛ (RCh) Choices: When a player lays down a conjunction (i.e. utters a formula with conjunction as its main connective), the choice as to which conjunct (left or right) is to be produced is yielded to the other player. The other player owns the choice as to which conjunct to request and, by repeating the attack, the challenging player may well end up requesting both conjuncts. This rule is readily justifiable by appealing to the extensional semantics for the conjunction of Propositional Logic: whoever asserts a conjunction should, if challenged, be ready to produce both conjuncts. A conjunction is true if and only if both conjuncts are true. For Ldlg, we should say: a conjunction is assertable if and only if the best strategy of attacking it gives to the challenger choice as to which conjunct to request.

When a player lays down an inclusive disjunction (i.e. utters a formula with inclusive disjunction as its main connective), the choice as to which disjunct (left or right) is to be produced is yielded to the player who utters the disjunctive assertion. The asserting player owns the choice as to which disjunct to produce when challenged. This rule is readily justifiable by appealing to the extensional semantics for the inclusive disjunction of Propositional Logic: whoever asserts an inclusive disjunction meets the burden of doing so by being ready to produce at least one of the disjuncts - it doesn't matter which one. An inclusive disjunction is true if and only if at least one of the disjuncts is true. For Ldlg, we should say: an inclusive disjunction is assertable if and only if the best strategy of defending it gives to the defender choice as to which disjunct to request when challenged.

VII.3 RULES OF CHOICES

X! φ • ψ			•1	
Y? φ or	Y? ψ	A_j	•2	[choice lies with Attacker]
↓	↓			
Y! φ	Y! ψ	D_j	•3	[Defender must respond by producing what is requested.]

X! φ ∨ ψ		∨1	
Y? ∨	A_j	∨2	[choice lies with defender]
↓			
X! φ or X! ψ	D_j	∨3	[defender chooses which disjunct to produce]

There is a readily intuitive justification of this type of rule. Whoever propounds or asserts a conjunction should be able to produce both conjuncts since a conjunction is true if and only if both of its component conjuncts are true. For an inclusive disjunction, however, the player who asserts the disjunctive statement carries her burden by being able to produce at least one - either one - of the disjuncts since truth of one disjuncts suffices for making the whole compound disjunction true.

VII.4 BRANCHING RULES

✇✇ (RBr) Branching: As in the Tree and Tableaux systems, certain states of the decision process or game induce branching: the two branches constitute two different sub-games of the game and both must close if the game is to be considered closed. In Ldlg it is the only the Opponent who can initiate branching under certain conditions (whose import can be understood by reflecting on the rules for choice presented above and by thinking of the Opponent as an Affirmer and the Proponent as a Negator or Denier if we want to refer to the branching rules in the Tree method.) The Opponent may or may not initiate branching: it is always assumed that each player is an ideally competent participant who implements the best available strategy for the purpose of winning.

Branching occurs: Briefly, when the Opponent attacks a conjunction or defends an inclusive disjunction (and material implication.)

a. when the Proponent has to defend a conjunction, as we know by now, the choice lies with Opponent who may initiate branching;

b. when the Opponent asserts/defends a disjunction, the choice lies with Opponent who may initiate branching;

c. when the Opponent asserts/defends a conditional: this is like the preceding rule b insofar as the propositional (material) conditional should be thought of as being essentially a disjunction:

$$(p \supset q) \equiv (\sim p \vee q)$$

VII.5 FIRST-ORDER DIALOGICAL LOGIC (LDLGPRED)

We extended the standard Propositional Logic idiom PROP to construct a formal system PRED for the standard predicate logic. Thinking of the quantifier symbols as denoting connectives (which we can do for stipulated finite domains), we would have to figure out what dialogical-logic rules we need for the two quantifier symbols. The rules will have to specify three rules – the trivial rule for stating a formula with the connective as main connective (which means, in this case, that the entire formula, omitting the quantifier symbol, lies within the quantifier symbol's scope); the second rule should specify how the quantifier symbol is to be attacked and the third rule should lay out the arrangements for defending against an attack. As we did above, we need to stipulate how the rules of choices constrain movements – which player, Attacker or Defender in a move, chooses which part of the formula to produce. Now, turning to quantifier symbols: we may think of the existential quantifier as an inclusive disjunction and we may think of the universal quantifier as a conjunction. With respect to choices: the instantiations of quantifiers yield formulas in which some individual constant symbol replaces uniformly occurrences of the variable controlled by the quantifier: this is the case for quantifier eliminations. The rule of choices in the case of dialogues for predicate logic will have to address the issue of choice of the individual constant. We will also need to apply the restriction that only the Opponent may make individual constants available; before that is done, such constants cannot be used by the Proponent. Which rule of propositional Dialogical Logic does this remind you of? What is the justification?

With respect to availability of atoms, the rule we need to lay down is that individual constants can only be made available by the Opponent. The Proponent can use individual constants only after they have been made available by the Opponent. This, clearly, has a significant impact when the Opponent has the option of enforcing a choice – as when attacking a universally quantified formula – and demands a specified individual constant; by specifying a constant that is not yet available, the Opponent as an attacking player can then win the game.

Rules for identity can be specified – we leave it as an exercise. We mention that negation of self-identity should be considered as unavailable for laying down and we see in the following example how this comes into play in a dialogical game for an invalid formula.

Player – signature	wff laid down	Action	Connective Rule for Move	Connective Rule Laid Down	Atom/ Constant Made Available
1. P!	Fa ⊃ ∃xFx	Th		⊃1	
2. O!	Fa	A1	⊃2		a
3. P!	∃xFx	D1	⊃3	∃1	
4. O?	∃/choice P	A2	∃2		

5. P!	Fa	D2	∃3		a is availabl e
	⊗ P wins	No moves Availabl e for O			

Hence,

Ldlgpred⊢ Fa ⊃ ∃xFx

We show now another example, in which branching is induced in accordance with the laws of this formalism.

Player – signature	**wff laid down**	**Action**	**Connective Rule for Move**	**Connective Rule Laid Down**	**Atom/ Con-stant Made Availab le**
1. P!	Raa ⊃ ∃x∃y(x ≠ y · Rxy)	Th		⊃1	
2. O!	Raa	A1	⊃2		a
3. P!	∃x∃y(x ≠ y · Rxy)	D1	⊃3	∃1	
4. O?	∃	A2	∃2 – P-choice		
5. P!	∃y(a ≠ y · Ray)	D2	∃2	∃1	a

6.	**O?**	∃		A3	∃2 – P-choice		
7.	**P!**	a ≠ a · Raa		D3	∃3	· 1	a
8.	**O?**	·		A4	· 2 -O-choice : Bran-ching		
		↙	↘				
9.	**P!**	No move for P	Raa ⊗				
		O wins branch	P wins branch				

Because P does not win both induced branches, O wins. This should not be a provable formula as it is invalid in its standard predicate interpretation. It turns out that it is not. The left branch leads to defeat for P because she cannot state an absurdity – which is the negation of self-identity. To the right, P wins because O has no subsequent opportunity for an attack, but this is not sufficient.

VII.6 EXAMPLES

1. Given this assortment of connective rules and strategic rules, let us see, by way of an example, how formula "p ⊃ (p ⊃ q)" is not provable - which ought to be the case since this is not a tautology of Propositional Logic. Opponent will win since the Proponent runs out of moves first, not being able to assert the consequent in response to Opponent's second attack.

Player – signature	wff laid down	Action	Connective Rule for Move	Connective Rule Laid Down	Atom Made Available
6. P!	p ⊃ (p ⊃ q)	Th		⊃1	
7. O!	p	A1	⊃2		p
8. P!	p ⊃ q	D1	⊃3	⊃1	
9. O!	p	A2	⊃2		
	⊗ **O wins**	No moves Available			**Atom q is not available**

Hence,

$\nvdash p \supset (p \supset q)$

Analysis:

1. Proponent initiates the game, dlg(p ⊃ (p ⊃ q))), by laying down the thesis, which is an implicative formula. The wff laid down is a thesis of Ldlg if and only if the competent proponent has a winning strategy for this game. The action in step 1 is laying down the Thesis. The major connective is material implication; hence the rule used for laying this wff down is ⊃1.

2. Opponent initiates an attack, the first attack of the game (A1), launched against the thesis. Opponent must use rule ⊃2 since the attack is against an implicative formula. Rule ⊃2 dictates that the antecedent of the implicative formula be laid down - if possible. Only the opponent may lay down an atomic variable letter for the first time, in accordance with one of the structural rules of Ldlg. The Opponent then lays down the antecedent, "p," and this atom, now conceded, becomes available to lay down by the Proponent as well.

3. The Proponent has to lay down the consequent of the wff that was attacked in accordance with the connective rule, ⊃3, about defending against attacks on conditionals. This the Proponent might not be able to do if the consequent is an atomic letter that has not yet been conceded by the Opponent; but in this case the consequent is not an atom and, hence, the Proponent, is able to lay it down as defense against A1 - hence, D1. The rule used is⊃3 guiding a defense against an attack on a conditional. The wff laid down by P is itself implicative: hence, the rule used for presenting the wff is ⊃1. This, of course, may invite an attack against the conditional in accordance with ⊃2.

4. The Opponent (O) attacks the wff laid down in the previous move in accordance with ⊃2; to do so, O lays down the antecedent, which is ⌜ p ⌝. This is the second attack of the game, A2, and, to be defended against, it requires the other party, the Proponent (P), to lay down the consequent, which is ⌜ q ⌝.

5. The atomic variable letter ⌜ q ⌝ has not yet been uttered by O. Hence, it is available to P to use in defending against the attack A2 and in accordance with ⊃3 (defense against attack on a conditional.) Nor does P have any other, delayed defensive or attacking, moves that can be undertaken by a proponent. Hence, there is no further move that is available and the game is terminated in accordance with the relevant structural rule. Since P ran out of moves, P loses - or O wins. This determines that the wff that was laid down in opening the game fails: it does not have the status of a thesis of Ldlg; interpreted for L, the corresponding propositional formula is not a tautology of that system.

2. p ⊢? p ∨ q

We must form the related implicative sequent -- turning the premise into the antecedent and the conclusion into the consequent.

⊢? p ⊃ (p ∨ q)

This is the sequent we check for tautology-status by application of Ldlg.

1. P! p ⊃ (p ∨ q)   Th ⊃1

2. O! p   A1 ⊃2 Atom

3. P! p ∨ q   D1 ⊃3 ∨1

4. O? ∨   A2 ∨2

5. P! p   D2 ∨3 P-choice

⊗ No moves for O; P wins.

No further available moves for O/ P wins

3. The classical tautology known as Expansion is not valid in the non-classical sublogic of Propositional Logic that is known as Intuitionistic Logic. It has been shown that there is no adequate truth-functional system whose interpretation matches the semantic motivations that underlie Intuitionistic Logic but a dialogical system enacted with the appropriate structural restrictions in place can give us exactly the Intuitionistic fragment of classical logic. Accordingly, by applying the restriction which, as we saw above, turns out to characterize the Intuitionistic system, we should be unable to prove the Expansion thesis (whose semantic interpretation is the classical tautology.) The restriction bans repetition of a defensive move already undertaken once by the Proponent. If this restriction is relaxed, the proof may continue and the best available strategy for the Proponent allows this role to win the game, thus establishing the credentials of Expansion as a thesis of the classical system. The availability of a modification in the structural rules, which permits capturing a logical language that has no matching extensional system, alerts us to an observation of far-reaching significance: structural aspects of systems were once thought to be extra-formal and otiose as opposed to the rules that govern the uses of the system's logical constants; nevertheless, the instance revealed by the constructability of Intuitionistic dialogues by means of structural modifications, shows both that structural rules play a form role in the definitions of logical systems and that the adjustments needed to proceed to intensionalist languages (which make Intuitionistic interpretations available) are themselves integral to the study of formal logic.

We juxtapose below the Intuitionistic and classical dialogues for the formula of Expansion. We respect different symbolic notations to indicate that we are dealing with different systems. It bears reminding ourselves that the dialogue are constructed in a metalanguage that is enhanced liberally with symbolic notations. We can think of the two dialogues below as importing into the dialogical metalanguage the symbols for the logical constants as {~, ∨, ⊃} for classical and {¬, ⊻, ->} for Intuitionistic Logic.

Classical	Intuitionistic
1. P! p ⊃ (q ∨ ~ q) Th ⊃1	1. P! p -> (q ⊻ ¬ q) Th ->1
2. O! p A1 ⊃2 Atom	2. O! p A1 ->2 Atom
3. P! q ∨ ~ q D1 ⊃3 ∨1	3. P! q ⊻ ¬ q D1 ->3 ⊻2
4. O? A2 ⊃3 ∨2	4. O? A2 ⊻2
5. P! ~ q D2 ∨3 P-choice ~1	5. P! ¬ q D2 ⊻3 ¬1
6. O! q A3 ~2 Atom ⊗no defense for ~2	
	6. O! q A3 ¬2 ⊗
7. P! q D2 Repeat! ∨3	⊗ - no Repeat for P; O wins.
⊗ No moves for O; P wins.	

4. We extended the standard Propositional Logic idiom PROP to construct a formal system PRED for the standard predicate logic. Thinking of the quantifier symbols as denoting connectives (which we can do for stipulated finite domains), we would have to figure out what dialogical-logic rules we need for the two quantifier symbols. The rules will have to specify three rules – the trivial rule for stating a formula with the connective as main connective (which means, in this case, that the entire formula, omitting the quantifier symbol, lies within the quantifier symbol's scope); the second rule should specify how the quantifier symbol is to be attacked and the third rule should lay out the

arrangements for defending against an attack. As we did above, we need to stipulate how the rules of choices constrain movements – which player, Attacker or Defender in a move, chooses which part of the formula to produce. Now, turning to quantifier symbols: we may think of the existential quantifier as an inclusive disjunction and we may think of the universal quantifier as a conjunction. With respect to choices: the instantiations of quantifiers yield formulas in which some individual constant symbol replaces uniformly occurrences of the variable controlled by the quantifier: this is the case for quantifier eliminations. The rule of choices in the case of dialogues for predicate logic will have to address the issue of choice of the individual constant. We will also need to apply the restriction that only the Opponent may make individual constants available; before that is done, such constants cannot be used by the Proponent. Which rule of propositional Dialogical Logic does this remind you of? What is the justification?

With respect to availability of atoms, the rule we need to lay down is that individual constants can only be made available by the Opponent. The Proponent can use individual constants only after they have been made available by the Opponent. This, clearly, has a significant impact when the Opponent has the option of enforcing a choice – as when attacking a universally quantified formula – and demands a specified individual constant; by specifying a constant that is not yet available, the Opponent as an attacking player can then win the game.

Rules for identity can be specified – we leave it as an exercise. We mention that negation of self-identity should be considered as unavailable for laying down and we see in the following example how this comes into play in a dialogical game for an invalid formula.

Player –	**wff**	**Action**	**Connective**	**Connective**	**Atom/**

signature		laid down		Rule for Move	Rule Laid Down	Constant Made Available
1.	P!	Fa ⊃ ∃xFx	Th		⊃1	
2.	O!	Fa	A1	⊃2		a
3.	P!	∃xFx	D1	⊃3	∃1	
4.	O?	∃/choiceP	A2	∃2		
5.	P!	Fa	D2	∃3		a is available
		⊗ P wins	No moves Available for O			

Hence,

⊢ Fa ⊃ ∃xFx

We show now another example, in which branching is induced in accordance with the laws of this formalism.

Player – signature		wff laid down	Action	Connective Rule for Move	Connective Rule Laid Down	Atom/ Constant Made Available
1.	P!	Raa ⊃ ∃x∃y(x ≠ y · Rxy)	Th		⊃1	
2.	O!	Raa	A1	⊃2		a
3.	P!	∃x∃y(x ≠ y · Rxy)	D1	⊃3	∃1	
4.	O?	∃	A2	∃2 – P-choice		
5.	P!	∃y(a ≠ y · Ray)	D2	∃2	∃1	a
6.	O?	∃	A3	∃2 – P-choice		
7.	P!	a ≠ a · Raa	D3	∃3	·1	a
8.	O?	·	A4	·2		

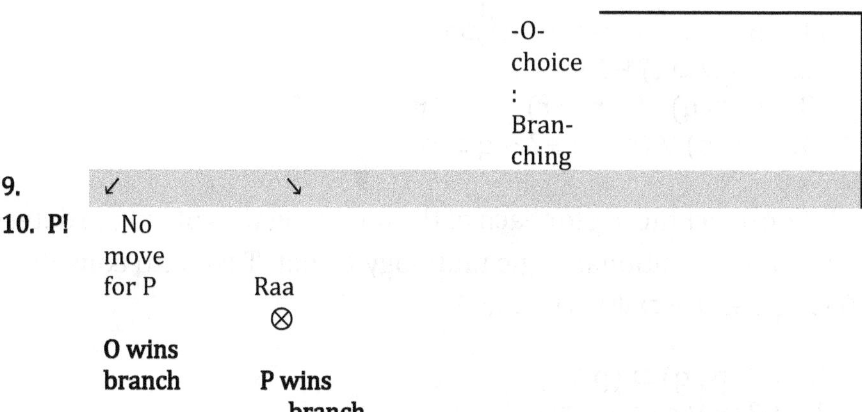

			-O- choice : Branching
9.	↙	↘	
10. P!	No move for P	Raa ⊗	
	O wins branch	**P wins branch**	

Because P does not win both induced branches, O wins. This should not be a provable formula as it is invalid in its standard predicate interpretation. It turns out that it is not. The left branch leads to defeat for P because she cannot state an absurdity – which is the negation of self-identity. To the right, P wins because O has no subsequent opportunity for an attack, but this is not sufficient.

VII.7 EXERCISES

(1) Check whether logical consequence obtains by means of the dialogical method: to determine whether it is or it is not the case that $\varphi \vdash \psi$ form the appropriate sequent $\varphi \supset \psi$ and construct the $dlg(\varphi \supset \psi)$.

1. $p \vee q \vdash ?\, p \cdot q$
2. $p \cdot \sim p \vdash ?\, q$
3. $p \vdash ?\, q \vee \sim q$
4. $(p \supset (q \cdot r)) \cdot (r \supset \sim s) \vdash ?\, p \supset \sim s$
5. $p \supset (q \supset r) \vdash ?\, (p \supset q) \supset (p \supset r)$
6. $(p \vee q) \cdot (\sim p \vee q) \vdash ?\, \sim q$
7. $\sim (p \supset q) \vdash ?\, q \supset p$
8. $\sim (w \vee t) \vdash ?\, \sim w \cdot \sim t$
9. $p \supset q \vdash ?\, (p \cdot s) \supset q$
10. $w \supset ((w \supset t) \cdot q) \vdash ?\, t \vee q$

11. $p \supset (q \supset r) \vdash ? (p \cdot q) \supset r$
12. $\sim (w \supset t) \vdash ? \sim t \vee s$
13. $(p \supset q) \cdot (\sim p \supset r) \vdash ? q \vee r$
14. $(p \cdot p) \vee (q \cdot q) \vdash ? \sim q \supset p$

(2) Construct the dlg for each of the following wffs of Ldlg to determine if it is a Propositional Logic tautology or not. Take into consideration: $(\varphi \equiv \psi) \equiv ((\varphi \supset \psi) \cdot (\psi \supset \varphi))$

1. $\vdash ? (p \cdot q) \supset (p \equiv q)$
2. $\vdash ? (p \vee q) \supset (p \equiv q)$
3. $\vdash ? ((p \vee \sim q) \cdot (\sim p \vee q)) \equiv \sim (p \equiv q)$
4. $\vdash ? ((p \cdot q) \vee (\sim p \cdot \sim q)) \supset (p \equiv q)$
5. $\vdash ? \sim (p \equiv q) \supset \sim (p \supset q)$
6. $\vdash ? (p \equiv q) \vee (q \equiv r) \vee (p \equiv r)$
7. $\vdash ? p \equiv ((\sim q \supset p) \cdot (q \supset p))$

(3) Construct dlg rules for the binary propositional connective symbolized by "$\subset$" and defined as:

$$p \subset q \stackrel{\text{def}}{=} \sim q \vee p.$$

Considering that the Proponent plays the role of Denier and the Opponent plays the role of Affirmer of posited wffs, refer to the tree rules for the binary propositional connectives denoted by the following symbols to design valid dialogical rules for the same connectives. (See IV.19 for the definable binary connectives of the standard Propositional Logic.)

$\equiv$

$\not\equiv$

(4) Based on the construction of PREDdlg, which we undertook above, continue by formulating rules-schemata for identity and for functions.

(5) Apply PREDdlg to generate dialogues for the exercises in chapters III and IV; you need to convert sequents to implicational formulas in order to produce dialogues and determine whether the sequents are or are not theses of the dialogical system.

INDEX

A

B

C

D

E

F

H

I

K

L

M

N

O

P

Q

R

S

T

U

V

W

Z

www.ingramcontent.com/pod-product-compliance
Lightning Source LLC
Chambersburg PA
CBHW070714160426
43192CB00009B/1188
* 9 7 8 1 8 4 8 9 0 2 6 7 1 *